PROJECT MANAGEMENT:
DESIGNING EFFECTIVE ORGANIS
IN CONSTRUCTION

PROJECT MANAGEMENT:

DESIGNING EFFECTIVE ORGANISATIONAL STRUCTURES IN CONSTRUCTION

DAVID R. MOORE

Manchester Centre for Civil and Construction Engineering

UMIST

Blackwell
Science

© 2002 Blackwell Science Ltd
a Blackwell Publishing company

Editorial Offices:
Osney Mead, Oxford OX2 0EL, UK
   Tel: +44 (0)1865 206206
Blackwell Science, Inc., 350 Main Street,
Malden, MA 02148-5018, USA
   Tel: +1 781 388 8250
Iowa State Press, a Blackwell Publishing
Company,
2121 State Avenue, Ames, Iowa 50014-8300,
USA
   Tel: +1 515 292 0140
Blackwell Science Asia Pty Ltd, 550
Swanston Street, Carlton South, Melbourne,
Victoria 3053, Australia
   Tel: +61 (0)3 9347 0300
Blackwell Wissenschafts Verlag,
Kurfürstendamm 57, 10707 Berlin, Germany
   Tel: +49 (0)30 32 79 060

A catalogue record for this title is available
from the British Library

ISBN 0-632-06393-9

Library of Congress
Cataloging-in-Publication Data
is available

Set in 10.5/12.5 pt Palatino
by Sparks Computer Solutions Ltd, Oxford
http://www.sparks.co.uk
Printed and bound in Great Britain by
MPG Books Ltd, Bodmin, Cornwall

For further information on
Blackwell Science, visit our website:
www.blackwell-science.com

CONTENTS

Contents

Contents

ACKNOWLEDGEMENTS

The author wishes to acknowledge the assistance of the following organisations and individuals:

- Rolls-Royce plc for its support in terms of both information and allowing the use of material developed from the Organisation Module of the MSc Project Management funded initially by Rolls-Royce and delivered by UMIST.
- Chen Hogbin, Division Chief CTGPC, for permission to include information on the Yangtze Three Gorges Project.
- The Association for Project Management (APM) for its permission to include the APM Glossary of Project Management Terms, and the work of Mr Mike Hongham in the development of the APM Glossary.
- Dr Mei-I Cheng and Dr Andrew Dainty of Loughborough University for their help on the subject of competency and competences that flowed from the EPSRC research programme in which we are collaborating.
- Mohamed Abadi for information on virtual teams.
- Brian Moore for information on Japanese woodworking techniques.

Damnan quod non intelligent
They condemn what they do not understand.

PREFACE

Aims and objectives

Unfortunately, it is rarely possible to achieve a situation where a book will contain absolutely everything that individuals would benefit from knowing about the subject(s) being studied. This is mainly because, in the manner of projects themselves, a body of subject knowledge is always undergoing change – new ideas are added, old ideas are adapted or discarded and, perhaps most frustratingly, no single author ever knows absolutely everything about their subject.

Rather than trying to list all knowledge concerning the subject of organisation structure in a project management context, this book seeks to achieve two aims:

- examination of the diversity of factors to be considered in the determination of an initial overall organisation structure for a given project; and
- examination of the possibilities for varying organisation structure in response to differing project environments over the lifetime of a given project.

These aims may seem typical of books currently available on the subject of project management, many of which discuss the problem of organisation. However, this book is not about the typical project management issues. It will not, for example, provide illustrations of how to optimise project durations through the use of critical path networks (CPNs) – there is plenty of coverage of such techniques elsewhere.

Rather than reinvent the wheel, this book concentrates on the task of encouraging a project to unfold in a manner which is as close as possible to that which can be identified as most favourable to achieving success. The means of doing this is through the designing and implementing of an optimum organisation structure for a project, and how the issue of organisation structure may make or break a project manager. A 'good' organisation structure will, as one example, be of significant benefit in achieving a project's optimum duration, whereas a 'bad' structure will be a significant hindrance. In neither case will there be any significant consideration of the number-crunching aspect of CPNs, PERT charts, and so on. There will, however, be consideration of how such techniques can be made aware of a project's structure requirements in the achieving of an optimised duration. In short, this book contains more discussion of organisation than is typical of a 'project management' book. It also contains quite a bit of discussion of some subjects that would not usually be expected in such a book.

Perhaps, then, the book is actually more relevant to the study of organisation (and organisations) than to project management. Again, there are many texts available that deal with the study of organisation and organisations, so there is little point in repeating much of what is already well covered elsewhere. There is no significant consideration, for example, of those over-arching needs of the majority of organisations to achieve immortality, grow in size, produce year-on-year improvements in performance, and so on. Quite the contrary: organisation is studied here from the perspective of creating something which has a definite lifetime, seeks not to grow beyond a certain size and aims to hit maximum performance as soon as possible, and then maintain it. In other words, this book deals solely with the study of organisation purely for projects and therefore falls into an area which lies between project management and the study of organisation – an area which has been poorly covered in the past.

A final point here is to identify the intended readership for the book. There is no intention to aim for the undergraduate-level market, as many of the concepts discussed are more appropriate to postgraduate-level study. Those who are currently functioning as project managers or have had recent experience of the role should find the book's content particularly useful and relevant, in that they will have encountered at least some of the points covered in a real-world context.

Objectives

The book has a number of objectives and these are grouped in the following manner:

- To provide tools for the identification of factors relevant to the development of an initial project organisation structure.
- To provide tools for the assessment of individual factors' significance in the operation of a project organisation structure.
- To advise on how the factors and their differing significances can be brought together to result in an initial project organisation structure.

These first three objectives can be regarded as a significant improvement on the traditional approach of simply imposing, for example, a matrix project organisation structure irrespective of what a specific project actually requires. These objectives are then followed by others:

- To provide techniques for identifying additional information requirements in order to further optimise the initial project organisation structure.
- To provide tools for the analysis of the additional information with regard to its impact on the initial structure.
- To advise on how to accommodate the additional information's impact on the initial structure so as to produce an intermediate structure.

These objectives should be viewed as an opportunity for the project manager to test the validity of the proposed organisation structure. There is the possibility of returning to the start of the process if any problems that emerge at this stage cannot be overcome. Essentially, a problem-prevention approach is being implemented rather than the more traditional problem-solving approach that is required when it is found that the imposed structure does not work. These objectives are then followed by these:

- To provide tools for identifying the extent of diversity for ways in which to implement the intermediate structure.
- To provide tools for the selection of the most relevant structure for each of the project's phases.
- To advise on achieving the project structure genome.

I hope that, after reading through the book, you will feel that all of the above objectives have been met. In order to place the various items covered within the book into a degree of order (but not control!), there are three main parts to the book. Each deals with the issue of organisation structure within a broad historical context. Part One, for example, deals with approaches to structuring project organisations over a period up to the recent past. Part Two concentrates on what may be regarded as current approaches and Part Three introduces a possible new approach. It is in Part Three that most of the book's objectives will be met.

PART I
STRUCTURE PAST

1

SEARCHING FOR THE REAL PROJECT: THE HISTORICAL APPROACH

Absus non tullit usum – misuse does not nullify proper use.

Introduction

Projects run on information. While it is true that information of itself cannot complete a project, in that it needs other resources to carry out the actual work, those resources can operate in a meaningful manner only when they are supplied with information. A bricklayer, for example, can lay bricks on the basis of his or her experience, but without project-relevant information there are many possibilities for error. These may range from minor errors, such as finishing the mortar joints in the wrong style, through to major errors such as building a wall to the wrong dimensions, in the wrong location, or using the wrong bricks. Information should allow processes to be performed effectively and efficiently. However, this is only the case when the information is relevant, complete and accurate. Information is therefore similar to the process by which it is distributed (communication) in that it can suffer from 'noise' – any factor which reduces its clarity and therefore its value. One of the points explored in later chapters (2 and 7) is the extent of an organisation structure's tolerance for information noise in comparison with the optimum tolerance for an individual project, but at this point it is sufficient to suggest that project organisation structures can play a key part in minimising or maximising noise with regard to the issue of project information.

This chapter examines what may be referred to as the traditional approach (which many organisations still seem to practise) in responding to noise as organisations attempt to define projects in terms of information. Some aspects of this approach have their roots in the mediaeval period, while others are (relatively) more contemporary, having emerged during the Industrial Revolution. A minority are positively cutting edge, insofar as they were developed during the second half of the 20th century. This chapter should be regarded as outlining the baseline from which many organisations will have to develop if the forecasts of ever more rapid rates of change during the 21st century prove to be correct. The content of this chapter should also be regarded as a deliberate attempt to raise more questions than answers. This suggestion may disappoint some readers, but possible answers, along with further questions, should become apparent as they read subsequent chapters. Expertise does not arise instantaneously – it tends to require some effort. Sorry!

1.1 Information as a production resource

During the mediaeval period, all of the major projects being carried out anywhere in the world were construction projects (wars could also be included, but there were just too many of them!), and certainly as far as Europe was concerned, the most complex of these projects dealt with two products: cathedrals and castles. There were no major motorways, hydro-electric dams, high-rise office blocks or complex petro-chemical industrial facilities to be built. Likewise, there were relatively few construction materials to trouble those working in the industry. Brick had largely fallen out of use with the decline of the Roman Empire and would not become a major material again in Europe until the Renaissance was under way. Plastics and other synthetic materials were unknown, and the most common metal in use seems to have been lead. Mediaeval constructors required only minimal performance information on a small number of materials: stone, timber, glass and so on. Consequently, the industry could take a relatively relaxed approach to the generation and distribution of information, and even large projects were structured to reflect this.

The organising force on many large projects during this period was the master mason, a highly skilled individual who could find himself rewarded for a successful project with considerable prestige and influence. Masons in continental Europe could rise to levels equal to minor royalty and there were recorded instances where the mason had sufficient power to take complete control of a project through deposing the client (Moore, 2001). The mason's authority was also supported by his guild, with its strict rules for progression through the recognition of skills and abilities. These skills did not include the production of working drawings as generated by a modern-day architect; there was frequently no definitive design at the start of construction as the process was more along the lines of a shared vision

4

between mason and client. In some cases the sharing of the vision was achieved through the use of a model of the intended structure. Such models could be significant projects themselves as they were sufficiently large to allow the client to walk around inside them. Perhaps they should be regarded as an early example of the current trend to produce CAD images that allow the viewer to 'fly through' the proposed structure. Certainly they could be regarded as prototypes of the intended structure.

Drawings were produced as the work proceeded, with individual masons and other members of the team producing pieces of the overall structure under the instruction of the master mason. The production of information could not therefore be regarded as a team effort in the manner of a contemporary large project, and this resulted in a structure that was largely concerned with the use of information rather than its production and control. That information which was produced related to the project objectives as determined by the master mason, and as the objectives did not seem to change considerably between individual cathedrals, or castles, due to the lack of innovation within the industry, there seems to have been considerable potential for the recycling of information between projects.

1.1.1 Project objectives

Objectives for a project can be a nightmare, as anyone who has worked on a project having frequently changing (or 'revised') objectives will know. In this regard the situation the mediaeval mason faced, with his authority to determine project objectives in a minimalistic manner, could well seem an ideal one. Nonetheless, objectives are an essential part of any project, for two reasons:

- they determine the resources required for the project and the manner in which they are to be used;
- they are the basis on which project success or failure should be determined.

Over time, project objectives have grown more detailed and demanding. The mediaeval objective to produce the largest cathedral in the country, for example, is a relatively simple one so long as there is an agreed means of comparing the newly completed cathedral with whichever cathedral was the largest at the start of the construction process. If the client then starts adding objectives, such as to complete the building within a given budget and/or within a specific time period, the project becomes more demanding. However, such objectives still have the saving grace of being truly objective in nature – time can be measured in the passing of days, and a budget in the spending of gold coin or whatever the currency may be.

By the Renaissance, new objectives were emerging that were essentially subjective in nature, such as producing the most beautiful cathedral in the land. Beauty, as the saying goes, is in the eye of the beholder, so what the mason may envisage as being beautiful, the client may well find ugly. This brought about a factor that contributed to the downfall of the master masons: innovation. Columns, for example, were deemed to be more beautiful when they were slimly proportioned, but the masons were used to constructing columns to achieve the objective of safety rather than that of attractiveness. Of course, there was no computer modelling at the time: each new building was essentially a prototype, and on each prototype dimensions were changed until cathedrals began to collapse, at which point the masons knew they had reached the limit of their material.

The gathering of expertise was therefore dependent upon a development programme based in part on the testing to destruction of prototypes. It would seem that, psychologically, the masons were not capable of designing buildings to go beyond the limits of safety that they had established. They were therefore unable to meet this new objective, unlike a new group who were able to produce such designs and who referred to themselves as architects. These architects were also able to produce good-quality representations of what the final building would look like and could therefore communicate their concept of beauty to the client so that there was a common basis for agreement. Essentially, these architects were able to overcome one example of information 'noise' because they were not constrained by having strongly embedded particular values into information, as with the historically established values of safety margins in the use of certain materials.

Unfortunately, the fact that these architects were unencumbered in their thinking by any significant performance knowledge concerning the materials they were incorporating in their designs also meant that they did not really know how to make them work on-site. For this, the masons were required, but they now found themselves increasingly working as subordinates to architects. Hence, the organisation structure changed as the need for innovation resulted in the development of a new specialism and the masons entered into a period of gradual decline during which they lost their prestige and authority. This situation raises the issue of being able to identify those characteristics that define relevant project objectives.

A relatively simple approach to the production of objectives is to treat them as success criteria: if all the objectives are met, the project is deemed a success. Such criteria have three desirable characteristics, which can be referred to as the 'three Rs' (although there are other variations in the literature):

- must be realistic;
- must be revisable (for when it is found that they were not realistic after all!);
- must be rational.

The need to be realistic may, at first, seem to be duplicated by the need to be rational, but there is a subtle difference between the two. Realism can be regarded as requiring that anyone developing objectives should regard the project in its true nature – take the project as it is rather than as you might desire it to be. Rational objectives are those based solely on reason rather than on emotion. While there are many instances during the completion of a project where it is desirable that those involved in the process become emotional about it (such as when a team spirit is required), there are dangers in allowing emotion to oust reason as the basis for project objectives. An example of this can be found in what is commonly referred to as Brunelleschi's Dome.

The good people of Florence wanted to build the most imposing cathedral in the land. Grand designs were duly drawn up and the final design called for a magnificent dome over one section of the building. This dome was bigger than any previously constructed in the region, which aroused Florentine pride. By the time the construction process had reached the base of the dome, however, Florentine pride was starting to turn into Florentine embarrassment as a different realisation dawned – construction techniques of the day would simply not allow for the building of such a large dome. The objective of constructing the largest dome had not been rational, and it was found not to be realistic. Unfortunately, it also proved not to be revisable, as the walls built to support the dome had taken considerable time and money to construct – repositioning them to allow the construction of a smaller dome was simply not an option. There would also have been the problem regarding loss of face.

Those in the congregation who ended up sitting in this area of the cathedral risked a soaking, as it was more than 100 years before Brunelleschi, a guild member in Florence with no apparent construction experience, encouraged by the offer of a prize to anyone who could come up with a realistic solution, finally found a way to construct the dome. Of course, it still took him 25 years to complete, but the result remains the world's highest and widest masonry dome (King 2000), a testament to Brunelleschi's ability to plan (just do not mention the prototype boat that sank while transporting 100 tons of fine white marble).

1.1.2 Planning to achieve objectives

Having ensured that all project objectives are rational, revisable and realistic, the planning process can begin in earnest. The focus of planning should be to achieve all stated objectives with the most efficient use of resources. A wide variety of planning techniques is available, from line-of-balance through PERT (probability evaluation review technique) to the production of simple bar charts – the possibilities can be bewildering. The project organisation structure should be capable of helping those within it to obtain the information they need to make selections from the possibilities available by identifying the true (rational) project as opposed to the mirage (emotional) project. Unfortunately, this is not always the case as projects fail due to the structure's inability (or unwillingness in severe cases of gatekeeping) to produce the required information, or its tendency to produce too much information (much of which may be erroneous) which then makes it difficult to extract that which is relevant.

An important factor with regard to planning is that of knowledge. There is a tendency, within the traditional approach, for some organisations to regard information and knowledge as being the same. As with the characteristics of rational and realistic, there are subtle differences between information and knowledge (as discussed further in Chapter 4). The latter can be regarded as being the range of a person's information, while the former can be regarded as being individual items within a person's knowledge. Straightforward? In order to clarify the matter further, consider the inability of discrete items of information to autonomously combine themselves with others; two plus two does not equal four because information is not itself able to carry out the addition function. Even at the genetic level, encoded information requires biological material to execute the addition. During the Industrial Revolution, there was much information available and as people were able to establish relationships between items of information, the extent of knowledge began to grow rapidly and the process of change in society accelerated.

Take, for example, the relationship between the machine that is claimed to have driven the Industrial Revolution, the Watt steam engine, and the problem of French cannons blowing up in the faces of their gunners. The Watt engine was a development of the earlier Newcomen engine. This was an interesting steam-powered engine, but was not sufficiently powerful to be useful as a versatile workhorse. Watt improved the design by realising that the main problem with it was the Newcomen's cylinder trying to do two things at once: be sufficiently hot to prevent the steam within it from condensing before it had moved the piston, while also being sufficiently cold to condense the steam as soon as it had moved the piston. By adding a separate condensing unit Watt was able to produce a much more powerful engine, capable of serious work. However, he also created a problem in that his new design required the piston to be a much closer fit in the cylinder than in the Newcomen engine, and there simply was not any way of producing such regularly precise pistons in 1769.

The problem remained until 1775 when a solution emerged as a consequence of those exploding French cannons. In 1773, M. de la Houlière, a French brigadier, arrived in England looking for someone to provide him with information on the production of safer cannon. He met up with an individual called John Wilkinson who, by the following year, had devised an entirely new method of boring cannons from solid castings. This new method was the first example of a guide being used in machine tools and enabled the boring of a cylinder that would not deviate by more than the 'thickness of an old shilling' (Burke 1978). Such accuracy was sufficient for Watt to begin production of the required pistons in 1775 and his engine became a huge success.

Only when a person (or, in certain cases, an artificial intelligence) comes into the picture does the potential arise for knowledge to be created through the seeking, or establishing, of such relationships between nominally discrete items of information. This establishment of relationships within a range of information is possibly the key skill exhibited by good planners and managers. They are also able to appreciate that objectives, as well as acting as success criteria, can act as constraints on the production process and innovation.

1.1.3 Objectives as constraints

Objectives can act as constraints in the obvious sense that if a 350-seater aeroplane is required, there is no point in constructing a four-seater biplane. While both may meet an ill-defined objective, such as the ability to transport people from one location to another, one of them fails dismally with regard to the scale of that ability. Perhaps unfortunately, project teams have a tendency to seek the reverse situation and turn a four-seater biplane into a 350-seater aeroplane. Well-defined objectives are essential in trying to reduce the possibilities for this sort of empire-building to occur (another example of trying to succeed in completing a mirage project), and in this sense are validly regarded as constraints. However, it is also possible to perceive an objective as having a negative impact on creativity, particularly if the project team is composed of individuals who would regard a 350-seater as a 'better' product than a four-seater, irrespective of what the client may think. The project structure would, in an ideal world, seek to avoid this perspective taking hold by fostering the perception of objectives as maximisers of creativity – in terms of balancing divergent and convergent thinking (Lawson 1986) – through the targeting of creative effort on relevant outcomes – a project team can be as creative as it desires, so long as its effort is focused on achieving those outcomes which are required for the project to be successful.

The focusing of creative effort on the achievement of relevant outcomes can be dependent on a number of factors, such as the leadership style of the project manager, constituency (membership) of the project team, nature of the project team life-cycle, and manner in which the project organisation structures itself. In this regard, possibly the most significant objective for the project manager is to achieve a project organisation structure which allows for creativity while also focusing that creativity. Many project managers fail to achieve this simply because their parent organisation has imposed its own inflexibility on the project through the predetermination of the project organisation structure as being the same as the parent organisation structure (the suggestion that organisations seek to replicate themselves by creating a twin external environment presents a possible exception to this argument and is discussed further in Chapters 4 and 6). Such an approach is evidence of a complete failure to recognise that parent organisations and project organisations have diametrically opposite perspectives on what may be referred to as the objective of life. Parent organisations are, by nature, seekers of longevity in that they desire to live as long as possible. The master masons possibly did not regard themselves as being parent organisations in that their emphasis seems to have been on the accruing and maintaining of prestige. In that sense, a parent organisation could be argued to have only existed around individual masons in two forms:

- the guild, which was composed of their brother masons and appears to have had a paternalistic character to its operation;

- those assistants and lesser masons who were immediately involved, under the master mason's instruction, with the running of individual projects.

The first of these forms could be considered an early example of what Handy (1999) refers to as a cluster organisation (see Chapter 4), but it is important to note that such structures do not engender a collective paradigm across the project team. However, the typical mediaeval project was run on the basis of authority and attached little significance to concepts such as collective paradigms. Project organisations should be structured to accept that the very nature of a project is that it will end, and that the end-point can be identified in advance and worked towards in an optimum manner. Such behaviour seems to have begun to emerge more clearly from some of the project-based organisations operating during the Industrial Revolution.

Along with the realisation that projects are faced with an ever-increasing rate of change in their external environment, it is perhaps tempting to argue that this difference of objectives between parent and project organisations with regard to longevity is becoming more and more blurred. After all, parent organisations do tend to re-invent themselves in response to significant changes in their operating environment, so at what point does one parent die and another is created?

The issue of differing longevity objectives is perhaps the most significant manner in which an objective can be regarded as a constraint, in that if the optimum project organisation structure is achieved, the issue of focusing creativity will be catered for by that structure. The project will have a higher probability of success and will therefore be able to contribute to a longer life for the parent organisation. There are, however, projects that are continuing to fail in this regard because they are structured on a contingency basis without any consideration of possible alternatives.

1.1.4 The contingency concept

Contingency is a concept which some project managers fail to fully appreciate and they therefore do not realise that they are adopting what is essentially a management approach based on it. In this regard they are no different from those who managed projects during the Industrial Revolution, but at least those managers had the excuse that nobody had identified any alternative forms of organisation. This issue of contingency will be dealt with more fully in Chapter 3, but as an introduction to the contingency approach, consider the following hypothetical situation (don't worry about the realism of the figures stated):

- You are asked to take on the role of project manager for a new project.
- The project duration has been calculated to be five years.
- The project budget is £50 million.
- The intended outcome from the project is 1000 miles (or 1600 kilometres) of electrified railway track.

Q: What would be the perfect organisation structure for this project?
A: There is no perfect organisation structure for any project!

Put briefly, the contingency approach seeks to achieve project success through controlling the risk of failure. The contingency mindset then becomes locked into finding its Holy Grail: one single perfect way to organise for all projects. However, by stepping back from this objective and viewing it rationally, it becomes apparent that it can never be achieved. This is because of the ever-present possibility of risk emerging within a project, where risk is seen as a relationship between the probability of something happening and the severity of the results if it should happen. Many project managers become fixated on this issue and will go to considerable efforts to produce contingency plans – what they intend to do if a particular risk does emerge. Some even go so far as to plan for secondary risk – the results flowing from the implementation of the contingency plan. Very much a case of 'If plan A fails, then work your way through plans B, C, D, etc.'. Within this approach there comes a point when the structure of the project begins to constrain the development of contingency plans, thereby precluding the implementation of alternative (and perhaps better) plans. The structure is then seen to be less than perfect. The problems of implementing this approach will be considered in more detail in Chapter 3 and in Chapter 6 where a different approach, emphasising project genomes rather than the issue of contingency, is introduced.

   As a consequence of the emphasis on contingency, the most appropriate (not perfect) organisation structure for a given project is generally selected from a limited number of possibilities. This situation arises because there is the inference that seeking to implement a new and innovative organisation structure will present a risk. One of the key factors in this selection process seems to be an awareness of factors such as the project's external environment, within which the project seeks firstly to embed itself, and secondly to survive, through developing an appropriate structure. In this regard, projects of any era have shared one consistent objective. The theory by which project managers have sought to achieve that objective has, however, changed over time. In order to deal with the concept of project environments, most project managers invoke (knowingly or not) a theory which first surfaced seriously around the mid-20th century: systems theory.

Memory test 1

At irregular intervals throughout the book you will find memory tests. These are not intended to test your understanding of the content to that point (there are other tests for that purpose) but simply to test your recollection of some key points within the book. In order to maximise their benefit you should attempt them from memory. You should then read back through the relevant sections and check that your recollection agrees with their content. This is important in that there are no separate answers given for these tests. In order to test your recollection to this point, please attempt the following questions:

(1) What were the two most complex types of construction projects in mediaeval Europe?
(2) Outline one example of the creation of knowledge from previously unconnected pieces of information during the Industrial Revolution.
(3) Give one example of a factor that can affect success in the focusing of creative effort on achieving relevant project outcomes.

1.2 Identifying the project nature in system terms

Miller and Rice (1970) provided one of the more detailed discussions of aspects of systems theory in relation to the issue of organisation. However, they were not the only researchers considering how systems theory might be of use in the 'real' world, and the area has continued to develop. Newcombe *et al.* (1990), for example, provided a useful definition of a system in terms of a group of inter-related entities focused on the performance of a function or reaching of a goal. Projects are clearly systems in that they bring together entities (anything with a distinct existence) and inter-relate them with the intention of achieving objectives. What may be less clear is the form of inter-relationship between those entities forming a specific project.

The nature of possible inter-relationships is particularly important when they are considered in terms of being both constrained by, and constraining of, the project organisation structure. If the approach of imposing a parent organisation structure on the project is taken, that structure may well constrain the possible inter-relationships that can be achieved and thereby potentially limit the effectiveness of the project in achieving its objectives. However, if optimum inter-relationships can be identified prior to developing a bespoke project organisation structure, that structure has the opportunity to be supportive of the project in achieving its objectives. In this sense, the project organisation structure is constrained (again, as with the issue of creativity, this is in a positive manner) by the identified inter-relationships.

In order to better understand the nature of this process, an overview of the main systems theory concepts is useful. For example, just what is a closed system?

1.2.1 Projects as closed systems

Closed systems have the main characteristic of being a group of inter-related entities where none of the entities, or the group as a whole, is capable of responding in a positive manner to changes in the external or internal environment. A typical example is a basic internal combustion engine. Once started it is incapable of responding to factors such as a diminishing supply of fuel (internal environment) or increasing air temperature (external environment) in any manner other than by ceasing to function. However, closed systems of this form, which may be referred to as being fully closed, can be difficult to find. In the case of the basic internal combustion engine, it is almost inevitable that anyone having the vision and determination to produce such a device would also soon realise the limitations of simply allowing it to run out of fuel.

A basic response would be to add some form of reserve supply that would be sufficient to allow the main supply to be topped up without the engine actually stopping. A more sophisticated response would be to add to the group of entities a further entity intended to provide advance warning of the fuel supply running out. Almost all vehicles these days have fuel gauges and/or warning lights to alert the driver that fuel will soon be needed. The engine then becomes capable of limited interaction with its external environment (through its operator's expertise in responding to its signals) and the system moves to being a partially open system. Such systems will be discussed further in a following section. Prior to that, the diametric opposite of a fully closed system needs to be considered: the fully open system.

1.2.2 Projects as open systems

Open systems are those capable of responding to changes in their external or internal environments. In this sense, you are a good example of an open system – when your internal environment tells you that you are hungry, you are capable of interacting with your external environment by going to the fridge and getting out the cheese/tomato/ lettuce, etc. (assuming that you remembered to interact with the supermarket part of your external environment). Unfortunately, even the human system has its limitations and when these are reached a human is just as dead as an internal combustion engine with no fuel. Fully open systems are therefore probably more difficult to find than fully closed systems. A regularly cited example of a fully open system is God, which should indicate just how versatile a fully open system can be.

Having established that there are two extremes of system, both of which are difficult to find realistic examples of, systems theory usually moves on to discuss intermediate forms of system which may be referred to as either partially open or partially closed.

1.2.3 The 'open–closed' system spectrum

At this point it is worthwhile considering further definitions of a system so as to be able to explore the concept of a closed–open spectrum in more detail. The internal combustion engine example considered the system in terms of a group of entities. Taking this further, it should be reasonable to view each of these entities as the result of a further system. The engine will require some form of piston, for example, and this piston will require to be produced (unless you are a fully open system and capable of thinking them into existence) by a system that requires the raw materials of pistons, along with some means of converting those materials into a piston of the required dimensions.

Systems can therefore also be viewed in terms of requiring inputs, such as the resources of labour, machinery and materials, which are then entered into a designed system of production where they undergo some pre-planned conversion to become the product of that system. A system can then be defined as anything which involves three factors:

- inputs of specific resources;
- a means of conversion of those resources; and
- the export of a planned product from the means of conversion.

This can be summarised as **I**(mport) **C**(onversion) **E**(xport) or ICE, a system definition which has been valid ever since humans first planned processes of production, probably centred on producing flint tools.

One useful aspect of this definition of a system is that it can also be applied to just about anything. Organisations, ranging from the smallest firm through to the largest multinational, can be seen as being systems of equal validity to that of an individual production system within a manufacturing facility. The ICE definition, or model, for a system is therefore potentially highly versatile and as such a powerful tool for analysing the requirements of an organisation, whether it be parent or project. This is particularly so with regard to identification of interfaces. However, it is not the only definition or model available. PAC (planning, analysis, control) is a further example of a TLA (three letter acronym!) system model.

1.2.4 Construction projects: predominantly open?

By applying the ICE model it is possible to consider a project system in terms of the extent to which it must inter-relate with its external environment, with the general heuristic being that the greater the extent of inter-relation, the more open the system will need to be. Mediaeval construction projects, with their relatively small range of materials and labour types, probably required less inter-relationship with their external environment than Industrial Revolution-era projects. Certainly, the speed of transportation was very slow in mediaeval England, as one example, so there was some incentive to make use of local materials that may have been present on site, particularly on large sites. The builders of Rhuddlan Castle in 13th-century Wales, for example, made use of the clay boulders that were available on site as rubble infill between the stone outer leaves of its curtain walls and towers (Wilson 1976). Just as a matter of interest, for those who may doubt the capability of mediaeval project management, Rhuddlan Castle was substantially completed in three years through the application of up to 3000 men (at a time when the population of London was around 35 000) in a remote area of the country, and the project was managed by one man, Master James of St George.

Other aspects of the Rhuddlan Castle project illustrate what can be regarded perhaps as a baseline for the openness of project systems. Project external environments are often defined in terms of four factors: political, economic, social and technological (PEST). The political factor certainly came into play on this project as the client was the English king, Edward I. Edward seems to have had no doubts concerning his overall requirements: he needed to subdue the Welsh rebels and decided that a key part of achieving this was to embark on a massive programme of castle building in Wales. This suggests that the real project was clearly apparent to Edward; this truly was empire building. Consequently, Master James enjoyed the full support of the king, but only so long as he was delivering the goods and was able to pull together the large workforce required, even though several other castles were being built at the same time.

Master James was effectively making positive use of the requirements and authority of the client and the project seems to have been organised with a structure which allowed him almost direct communication with the king. However, the project did not need to be organised so as to consider the present-day diversity of political forces such as health and safety legislation, building control, building regulations, etc. In other words, the project did not need to be as open to political factors (in terms of diversity) as a modern project would.

Another factor for consideration is the technological one. Wilson (1976) suggests that castles were simply architectural propaganda in that they were built with the intention of dominating the people within their district. In order to achieve this, castles had to be both impressive and impregnable, requirements which placed considerable emphasis on the technology available for their construction. This technology had evolved in England since the arrival of the Normans, with the added input of castle technology from the Middle East. One feature of this evolution that can be identified among the series of castles of which Rhuddlan was one was the movement to the concentric wall layout. This allowed the occupants to rain down concentrated fire power from several levels and reached the peak of its development in Beaumaris Castle, which took more than 30 years to complete.

Modern construction projects have been forced to interact more widely with their external environment through the emergence of factors such as an ever-increasing range of materials which are not available as a natural resource on the site itself, and the greater extent of specialisation within the labour force. This latter factor can be traced back to the Industrial Revolution and the emergence of mechanised production processes in the manufacturing industries that powered it. If the manufacturing industry was a factor in the construction industry developing more open systems, what is the extent of openness in manufacturing's projects?

1.2.5 Manufacturing projects: predominantly closed?

At this point we reach one of those problems that result from standard definitions. The standard industry classification (SIC) used in the UK to define the activities of the construction industry is very wide ranging and includes some nominally manufacturing activities. It also includes activities such as demolition, quarrying, and erection of overhead cables. Combine this with the definition of a project as anything with a start, middle and end, and the potential for true projects in the manufacturing industry is reduced – the emphasis within manufacturing activity tends to be on achieving continuous production. Producing one Reliant Robin (don't worry if you do not know what one of these is, some would argue that you are lucky!) is not the objective. The objective is to produce as many as the market can take. Much manufacturing activity can therefore be argued to be in the area of general management; concerned with managing the *status quo* rather than managing the process of change.

The areas within manufacturing concerned with managing change are essentially research and development (R&D) projects and those small manufacturers which produce individual (or very short production run) bespoke items. Organisations producing several million widgets per day are of little relevance to this text. Consider the example of Mr Dyson of Dyson vacuum cleaners. The story goes that the development period of his cleaner covered many years and several hundred mock-ups and working prototypes. Only when he was fully satisfied that he had a product meeting his exacting specification did the R&D project come to an end. This is not to say that R&D stopped completely – several new versions of the cleaner have been produced since commercial production began, but those versions have been produced in the context of environments that varied from that of the initial R&D project. Consequently, they will have been project managed within a different organisation structure, even if it is different only because Mr Dyson can now afford to hire some help to carry out the R&D.

The temptation is to reinforce the old stereotype (a manifestation of the social aspect of the external environment – see the earlier discussion of PEST) of the manufacturing industry as being composed of noisy, dirty machines thrashing away furiously in pursuit of the Holy

Grail of producing vast quantities of a product in the shortest possible time. However, many of the processes and products spawned as the manufacturing industry grew during the Industrial Revolution have long since gone from the UK scene, and indeed from much of the developed world. Modern manufacturing has changed in at least one significant respect – it appreciates that the rate of change in the demand for products has, and is continuing to, accelerate.

This is perhaps nowhere more apparent than in the area of motorcycle production. The great British motorcycle industry at its peak ruled the world of powered two-wheelers (PTWs) and it could get away with producing almost any old rubbish – the demand from its external environment was so huge that almost anything with a Triumph, BSA, Norton or whatever badge on it would sell irrespective of how unreliable it was. Some manufacturers rode this wave with barely any consideration for possible external environment change, such as in the demand from their customers. Phelon & Moore, for example, produced one model of its Panther motorcycle which underwent little change in around 60 years of production (McDiarmid 1997).

In summary, the manufacturing industry as a whole is becoming more of an open system, but it is not a project-based industry in the manner of the construction industry. However, one of the areas in which it is arguably more receptive to its external environment is the use of information technology (IT), particularly for the use of computers to carry out analyses.

1.2.6 IT projects: completely open?

The term 'computer' did not originally apply to machines – the first use of the term seems to have been in reference to the specialists who would manually carry out the long and complex calculations required to model public health projects, such as the Victorian period London sewer system. These specialists would work in teams, with each member of the team specialising in one part of the overall calculation (Petroski 1996). This sort of specialisation was typical of the Industrial Revolution, but it seems somewhat ironic that what was once a highly specialised human (having very little interaction with its external environment) should have evolved to become a machine which is increasingly capable of processing vast amounts of data concerning both internal and external project environments. But does this ability contribute towards making IT projects more open than other projects?

It is perhaps arguable that the IT industry runs projects which can suffer from being too open. Consider, for example, the so-called Moore's Law – computing power doubles every 18 months. Not so long ago a computer's memory capacity was measured in kilobytes, but memory capacity has grown to the extent that it is now measured in gigabytes. In order to be able to realistically make use of such a huge memory, the information within it needs to be accessed rapidly, otherwise we are back to the problem that resulted in those human computers, and so processing speeds have risen proportionately. The average home computer now has more power than the Apollo lunar landers had.

Such power presents many possibilities, such as the organisation within the SETI (search for extraterrestrial intelligence) programme of a project to harness several thousand PCs to process information which would otherwise have required a rather expensive supercomputer. Such a project had to be very much towards the open end of the system spectrum for no other reason than that the majority of the resources were outside the parent organisation and could be accessed only when there was computing power to spare – information was processed between the PCs as and when they had nothing else to do. Managing several thousand pieces of information that may all have been processed to different levels by computers dotted around the globe, and then putting together all the results of that processing, would seem to require a highly open system of project organisation.

It could be suggested, then, that the extent of a project's openness to its external environment can have considerable effects on its internal environment and system of organisation. A point worth further consideration is whether there are circumstances in which the external environment can force itself onto the project's internal environment, irrespective of the project's extent of openness?

1.3 Effects on the internal environment

One of the claims for considering a project in terms of its internal environment is that this allows a boundary to be placed between those areas outside the control of the project organisation and those areas that can be controlled by it. There is of course an issue with regard to the extent of control that can realistically be achieved within the internal environment, and this will be examined in more detail in Chapter 5. At this point the emphasis is on the control issue from the perspective of considering situations where the external environment can negate the control characteristic of a project's internal environment. By proceeding in this manner, an awareness of factors concerning the issue of project organisation structure flexibility can begin to be raised.

1.3.1 Ethics as an environmental force

Philosophers may grandly state that ethics are a set of moral principles by which we all should be guided in the eternal struggle between good and evil and indeed they would be correct in doing so. Unfortunately perhaps, the majority of the human race, while not being evil, give the impression of not being quite so fervently wedded to idea of living within the confines of any absolute set of principles. Many of us seem to prefer a rather more fluid approach to the matter of what is good and what is not so good. After all, we have moved on as a society from that of Master James of St. George in which the majority of people would blame a building collapse on evil demons and invoke various superstitions. We now realise that buildings collapse either because of overwhelming natural forces, such as earthquakes, or more commonly because one or more of the people involved in the construction got something wrong, deliberately or otherwise.

Perhaps the error was in the form of an inaccurate structural calculation, or they may have cut the occasional (or frequent) corner with regard to the quality of work and/or materials used. The mediaeval masons, among others, recognised the possibility of such temptation arising and were keen to reduce it through the adoption of a code of conduct or set of professional ethics. Such an approach continues today, with all the professional bodies imposing some form of code of conduct on their members. These codes should be seen as being sets of professional, rather than societal, ethics in that they usually go beyond the moral principles of right and wrong found in societal ethics. Carey and Doherty (1968) asserted that professional ethics may be regarded as a complex mix of moral and practical concepts. The practical aspects of professional ethics tend to be focused on comforting the society which provides the customers for a given profession's members. After all, would you be willing to undergo brain surgery at the hands of someone who was not a member of a professional institution with a stated aim to preserve life and cause no harm?

As far as the project's internal environment is concerned, the ethics import can become somewhat complicated in that it should comprise two components: an individual's understanding of their society's ethical beliefs, and that same individual's understanding of their professional (or trade) body's requirements regarding professional ethics. However, while any member of such an institution may 'talk the talk', they do not always 'walk the walk'.

Of course, some organisations take the potential for such a situation arising more seriously than others. This may be a reflection of the values and beliefs of the organisation, or it may simply be an awareness of the nature of the society in which that organisation exists. At this point there is no intention to explore either of these possibilities in any great detail. Rather, the issue is mentioned as one of potential relevance to an increasing number of organisations. Luther (2000), for example, notes that the number of US companies using integrity testing has increased to around 5000 and that approximately 2.5 million people are tested each year as part of the personnel selection procedure. Luther also notes that such tests are increasingly being seen as valid indicators of job performance and counterproductive (dishonest) behaviour.

Fan *et al.* (2001) suggest that quantity surveyors in Hong Kong are more pessimistic and confused with regard to their professional ethics than is the case with accountants. While Fan *et al.* consider a number of possibilities as contributors to this situation, it is particularly interesting to note that around twice as many (34%) accountants attended ethics-related courses as part of their college/university course than did quantity surveyors. The educational emphasis within the professional ethics area for quantity surveyors in Hong Kong seems to be around 80% on practical concepts and 20% on moral concepts.

The issue of emphasis within the professional ethics mixture of concepts leads to the suggestion that perhaps construction projects are largely reliant upon their human resources having developed a moral perspective as part of their socialisation. If so, two key factors in issues such as integrity testing would seem to be:

- Does the individual(s) internal to a project accept the values and beliefs of the society external to the project?
- Are those values and beliefs ones which may be viewed by the project organisation as being positive in nature, such as a belief in not removing anything from the project which is someone else's property? Not taking home pencils from the office, for example.

If both these factors are present, society can then be seen to be operating as an external constraint on activities forming the project's internal environment. However, it can be argued to be doing so in a less precise manner than the additional external constraint of legislation in which the act of theft is well defined in comparison to a more general concern with not removing someone else's property. After all, if you cannot identify who owns an individual item of property, it is a small step to say that it belongs to no one and can therefore be removed legitimately.

As an example to illustrate the possibility of differing levels of what may be referred to as constraint precision, consider the societal constraint of not killing others, and the legal constraint of the COSHH (control of substances hazardous to health) regulations enacted in the UK. People can be killed in a variety of ways, and as this is not a textbook on euthanasia, that variety will not be discussed here. Instead, the emphasis is on the differences of perception regarding killing directly, or murder (the intentional killing of one human by another), and indirectly, or manslaughter (the unintentional but not accidental killing of one human by another). The societal constraint may regard murder as completely unacceptable, while regarding manslaughter more ambiguously: it's not as though she *intended* to kill him, is it? If everyone joining the project from the external environment (society) has such beliefs, the project environment should be a pretty safe place to be.

However, construction sites are generally among the most dangerous places to work. A factor in this situation is that of hazard perception, particularly when the hazard applies to others rather than oneself. If the project's internal environment does not emphasise hazards in a negative manner (i.e. avoid them), people may be killed or injured unintentionally, but not accidentally in as much as the hazard was recognised but not responded to in a positive manner. Construction workers are one of those groups who have a stereotype as risk-takers: what do they care about a piddling little hazard such as using paint with a high level of volatile organic compounds (VOCs) in a room with insufficient ventilation? After all, what harm did a strong smell ever do anybody? The wrong kind of strong smell can do considerable harm, and this is where COSHH comes in.

COSHH seeks, among other things, to ensure the provision of information on hazards resulting from substances that may be used on construction sites and elsewhere. Such information can be regarded as an import to the project internal environment, in that it should result in a culture of hazard elimination or, where this is not possible, reduction. When viewed in this manner, the COSHH regulations have a more specific outcome than the societal constraint of not killing anyone directly, in that they cause the project organisation to take steps to minimise the potential for killing anyone indirectly through the use of hazardous substances. These regulations become even more constraining when considered in conjunction with other legislation that can be regarded as supporting them. An example here would be the Construction (Design and Management) Regulations 1994, generally known as the CDM regs, with their requirement to minimise risk through a range of strategies such as the replacement of a hazardous material with a less hazardous one. By having one regulation to force the supply of information, a further regulation can then bring about actions based upon that information in quite a precise manner.

In considering ethics as an environmental force, it becomes evident that some imports are complex in nature due to their being inextricably linked with other possible influences such as the legal and societal imports. Project organisation structures have therefore had to increasingly respond to the unavoidable fact that a desired import to their internal environments can also bring along undesired imports.

1.3.2 From the external to the internal: system imports

The consideration of desirable and undesirable imports raises a further problem for project organisation structures, and this relates to the issue of control. System models other than ICE can be found in the literature, one example being PAC, which looks at systems in terms of the functions essential for the creation and maintenance of a system. In the PAC model, control is an essential function, and this can also be stressed in the ICE model in that if a required export is to be achieved, the import and conversion activities will have to be controlled or regulated. A further consideration within this issue is the assertion that production is not accidental (Moore & Hague 1999): it must be capable of replication, and in the traditional model this raises the need to be able to analyse a process, plan for its execution and then control it when it is implemented. One means of allowing for this has flowed from the concept of system boundaries, in particular the concept of import and export boundaries.

Miller and Rice (1970) concluded that, for a system to work effectively, it needed to be capable of regulation and maintenance. The latter will be picked up in more detail in Chapter 2, but regulation is worth further examination at this point. In the previous discussion of closed systems it was noted that they were typically not capable of responding to their external environment. However, it is important to appreciate that this does not mean that they are not capable of regulation. It may even be argued that a closed system needs to place more emphasis on regulatory activities (which can also be viewed as subsystems within the main system) than an open system.

Consider the previous closed system example of a basic internal combustion engine. An internal combustion engine relies, in essence, on the so-called triangle of fire: fuel, ignition source and oxygen – if any one of these is missing, combustion will not take place. This then suggests at least three subsystems: fuel delivery; ignition source and exhausting of the results of combustion. Each of these will need to be regulated so as to occur in the correct relationship to each other. There is little use in having an ignition subsystem providing a spark prior to the fuel delivery subsystem having provided the fuel to be ignited. Likewise, the project organisation structure must be capable of regulating the required imports in terms of at least three criteria: time, cost and quality.

A feature contributing towards achieving this form of regulation is that of the project import boundary. This is set up to enable the rejection of any imports which arrive too early or too late, in too large or too small a quantity, or of an inferior quality (imports of a quality superior to that required can also be rejected, but this is seldom a problem!). Of course, the project organisation structure must also be capable of communicating to individual suppliers of imports the relevant information regarding time, cost and quality.

With regard to quality, for example, around 2000 years ago the Roman architect Vitruvius sought to provide guidance for his fellow architects by identifying quarries in central Italy capable of supplying stone of excellent quality. However, if an architect chose to use stone from a quarry nearer to Rome, Vitruvius recommended that such lesser-quality stone be weathered outside for two years so that any imperfections in the stone would result in its cracking. Such cracked stone was still too valuable to be wasted and Vitruvius suggested it should be used in foundation construction (Hill 1996).

If a given project was to follow this guidance, it would need an organisation structure capable of making a decision regarding the location from which stone was to be extracted and, if the inferior quarries were to be used, planning the related stonework activities to commence at least two years after the stone was extracted.

Figure 1.1 suggests one representation of this relationship between boundaries and imports. It also includes the relationship between boundaries and exports.

1.3.3 From the internal to the external: system exports

Given that considerable effort may go into the planning of a project, it would seem only reasonable that such effort is targeted at the achieve-

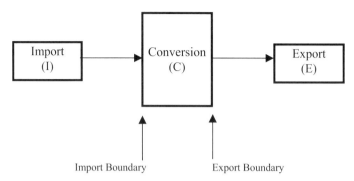

Fig. 1.1   ICE system model.

ment of the desired system export. This may be a building, ship, car or vacuum cleaner. However, it has long been recognised that specifying the required export poses particular problems regarding information. A good example of this problem can be found in the 13th-century statement from the Commune of Florence that the proposed cathedral of Santa Maria del Fiore, when completed, had to be more beautiful and honourable than any temple in any other part of Tuscany (King 2000). How, then, were these criteria to be determined as having been met in full? The problem was resolved to some extent as the work on the cathedral proceeded, and by the time work on the grand dome was due to be started, a 14-page specification for its construction had been produced. This approach of using a written specification has continued to develop and many projects now routinely produce specifications running into hundreds of pages in which reference is made to factors such as standards and codes of practice.

The standards-based approach to specifications is particularly useful in that it can largely determine the boundaries of a project through what is essentially a definition of the conversion process: what is to be done, how it is to be done, and the standard to which it is to be completed. The desired export is defined in terms of everything that goes into its completion; all the project manager has to do is make sure that everything comes together in the correct order. Seems deceptively easy until consideration is given to whether or not this approach truly defines the real project. It may well be that it defines a process of production, but projects tend to be about more than just a production process – if they were not, then the only form of management would be production management as project management would not be required.

A further complication is that, unfortunately, the conversion process has the potential to result in many exports in addition to the desired (and hopefully, accurately defined) export. In the current environment of worrying about making production processes sustainable, the export of pollution, as one example, is increasingly becoming regarded as undesirable. The conversion process within any project, past, present or future, can therefore be viewed as evolving over time (in response to the emergence of new technologies and materials, for example) while being constrained by both the limitations of the imports and the expectations of the export(s). In this sense, the real project also changes and it may in fact be the product exported from the project that becomes the mirage. Something to think about for a while before you attempt the next memory test!

Memory test 2

    (1) Professional ethics may be regarded as a complex mix of ...?

    (2) What unavoidable fact do project organisation structures have to increasingly respond to with regard to the import of a desired import to their internal environment?

    (3) To which essential ability of a system is the concept of an import boundary relevant?

    (4) What problem in the context of specifying the required export was the cathedral of Santa Maria del Fiore suggested as an example of?

    (5) In what way is the standards-based approach to specification writing particularly useful?

1.4 The conversion process

Put simply, the conversion process is the point in a system where the actual production takes place – all other aspects of the system should be seeking to support this. Problems begin to emerge as the conversion process grows in extent and complexity, and these two factors are potentially linked through the characteristic of production which can be referred to as functional specialism. Since the beginning of the Industrial Revolution both the degree of specialism and the range of specialists within the workforce have increased. Within UK construction this trend has resulted in a highly fragmented industry, a situation that has had follow-on effects with regard to planning and control so as to ensure that the required fragments come together (integration) as and when needed. This process (functional specialism) has been driven by researchers such as Frank Gilbreth, the father of scientific management. It was Gilbreth who invented the 'therblig' (an anagram of Gilbreth), or micro-movement, for use in work study. This system has seen further development since its inception, with one new therblig being added to the 17 originally identified by Gilbreth.

The therblig illustrates the extent to which researchers, and others, have sought to improve any process of production involving human resources. Therbligs are micro-motions of the hand and/or eye that can be used to describe a system of work. They can also be used to construct a new system of work from scratch if desired, and amend existing systems by seeking to remove ineffective elements of movement. However, such micro-detail is of little relevance within this consideration of project organisation systems beyond its effect of making planners identify specialist tasks within the conversion process they are seeking to implement, and this point will be developed further in Chapter 2. This identification can be done on the basis of locating points of differentiation within the conversion process. Such points occur at the boundary between two tasks: each task then becomes a system (or more accurately a subsystem) in its own right and can be referred to as a task subsystem.

1.4.1 Task systems (subsystems)

For the early background to the concept of task systems, we need to return to the work of Miller and Rice. Their perspective on task systems was to view them in terms of the three 't's: time, technology and territory (Miller & Rice 1970). The suggestion was that if any significant difference in any of the three 't's within a system could be identified, this was evidence of differentiation and thereby the single system was effectively at least two systems. As with any theory, there is some scope for interpretation within this one. The concepts of time and technology are, it is hoped, fairly well established and understood by most project managers, but the concept of territory is possibly generally less well understood.

Moore and Hague (1999) gave one interpretation of how the concept of territory may be interpreted in terms of both physical location and knowledge and skills. Dealing with the aspect of physical location first, it should not be too difficult to view production processes happening at opposite ends of a project site as being differentiated in terms of territory but still being part of the whole project. Two separate processes, each producing 50 units of the same output in a given time, cannot be regarded as being the same as one process producing 100 units of output in the same time. If nothing else, each of the separate processes will require its own supply of imports and this raises issues of planning and control. The processes are therefore individual subsystems within a project. So, while the two processes may be identical in terms of the imports required (materials, equipment, skills, etc.), they are classed as separate subsystems if there is a significant difference in their physical location on the project site.

Moving on to the aspect of knowledge and skills results in a slightly more complex picture emerging. The first problem is that of regarding knowledge and skills as being capable of existing separately from the technologies of production. Technology is a possible area of differentiation in its own right, as previously mentioned, so when there is a significant change in production machinery, for example, it would be fair to decide that differentiation has been identified. However, what if there is no change in the production machinery, but the export of the system can be made to be different purely on the basis of the machine operator's knowledge and skills? This would suggest that a change in the knowledge used results in a new system – the individual involved has moved on to a different location in their knowledge territory.

There is, of course, always the possibility of arguing that if one operator starts off producing one type of export and then moves on to a second type, there is actually a differentiation in terms of time and there is no need to consider knowledge territory at all. This argument has a number of attractions to it, but unfortunately it also raises its own problems – if time is an import, then knowledge and skill must also be imports, so to take an approach of considering differentiation only in terms of time and technology results in a significant aspect of the production process being placed outside the functions of planning, analysis and control. In other words, knowledge and skill would be placed outside the system when we all know (yes, you do) that without these factors production processes just would not happen. Such an approach would also remove a useful tool in helping to identify just how many subsystems a project system requires.

1.4.2 How many subsystems?

As the concept of subsystems begins to take hold, the question emerges of how many of them are required by a project. The simplistic answer to this question is that all projects require the minimum number with which they can achieve the required export. As the old management mantra goes: 'keep it simple, stupid.' Unfortunately, this mantra can be taken too literally and if simplicity becomes the overwhelming objective, the quality of the output from projects is likely to diminish.

Consider the jointing of timber sections, for example. This can be achieved quite simply by driving a large metal spike between the two or more sections to be joined. Not particularly elegant, but if simplicity is the sole objective, the solution is acceptable. Not only is the solution possibly inelegant, it is also constraining in a number of ways, such as the need to have available a supply of metal spikes. It is also constraining in terms of technological development in that if the supply of spikes dries up, production of joined timber units will have to stop as nobody knows of an alternative method. A final considera-tion is that the method is not really making use of the properties of the timber, so there is no need to be aware of such properties and again, technological development does not happen.

As an example of technological development in the use of timber, Japanese carpentry takes some beating. Because the Japanese put ef-fort into understanding both the properties of timber and how tim-ber may be worked, outstanding examples of elegant joinery (Sato, Nakahara, 1967) have been incorporated into many Japanese historic buildings. Admittedly, they do seem complex, but there is not a metal spike in sight. The suggestion is that, while the efficiency of the project can be viewed solely in terms of using the absolute minimum of sub-systems, the project also needs the flexibility not only to incorporate new ideas but also to develop a few of its own. In other words, it needs to allow for creativity to emerge during the project's lifetime.

Once a realistic number of subsystems has been identified, there comes the moment of truth: can they be structured in such a way as to provide the required export within any stated constraints of time, cost and quality? In order to do this, there needs to be some considera-tion of the role of precedence.

1.4.3 The role of precedence

Precedence can simply be viewed as being a representation of the de-pendencies between individual subsystems within a project. Many of these are historical in nature, one example being that foundations are generally constructed before the walls are built. Modern technologies mean that this sort of precedence relationship is no longer the abso-lute given that it may have been previously. It is now quite possible to devise a form of construction that breaks this relationship, and there may be instances in which such a break is of benefit. However, revis-ing well-established relationships within production processes can be a fraught process and is generally best avoided unless clear benefits from the process can be identified.

The relationship can also be advised by the consideration of differentiation, in that the nature of the differentiation may suggest the most appropriate precedence relationship between the resultant subsystems. One of the subsystems may well depend upon the export from a previous subsystem before it can begin its own production process. For anyone wishing to examine the role of precedence further, reference to any good management book dealing with critical path methods (CPM) is suggested.

1.5 Conclusions

This chapter has outlined historical aspects of problems in identifying the real, as opposed to a mirage, project. Such problems continue, and have arguably worsened due to the increasing complexity and range of demands placed on modern projects. While project managers from past eras would identify with the old and true demands of bringing in a project within budget, on time and to quality, many of the current constraints imposed, along with the range of technologies available, would be completely unrecognisable to them. There is no suggestion that this is a bad thing – if Master James of St George could truly be dropped straight into the role of project manager on the modern-day equivalent of Rhuddlan Castle without experiencing any problems, then project management would indeed still be in the 13th century. Obviously this is not the case. Change is inevitable, and project managers should be more aware of this dictum than many other professionals if for no other reason than that nobody has built a Rhuddlan Castle for a very long time.

An important response to change is the ability of a project organisation to structure itself to be able to deal with whatever change manifests itself, and one of the relevant factors introduced in this chapter is that of treating projects as if they are open systems. This has not always been the case, but as the rate of change faced by projects (particularly those of long duration) has accelerated, the open system concept has become increasingly relevant. However, there remains a varying level of uncertainty for project managers as to how the theory of open systems can be applied to real industrial projects. Only when project managers are more certain in this regard can they begin to consider how to optimise their project organisation structures.

Chapter 2 will examine some of the more relevant open system characteristics and seek to apply them to industry. Chapter 3 will then examine some of the implications of failing to focus on the relevant environmental forces within the open system model, before a possible model for designing organisation structures is examined (finally!) in Chapter 4, by which point you will be around halfway through the book.

2 OPEN SYSTEM CHARACTERISTICS IN INDUSTRIAL TERMS: RELATIVELY RECENT DEVELOPMENTS

Cessante causa cessat et effectus – when the cause is removed, the effect disappears.

Introduction

This chapter will seek to examine the problem of defining the real, as opposed to the mirage, project through the further consideration of open system characteristics. Because systems theory is a relatively new body of thought, this chapter will also introduce some of the more exotic ideas concerning the operation of systems. One example of this is the issue of entropy and how it may affect a system. The purpose in adopting this approach is to encourage the thinking process to step up a gear and become more creative – an ability that project management environments, unfortunately, do not always encourage, even when they claim to do so.

Creativity can be regarded as being the balancing of convergent and divergent solutions to a problem (Lawson 1986) and if this is not recognised, much energy may be wasted by the project system as it tries valiantly to impose an inappropriate solution to an ill-defined problem. Within technology-based industries, perhaps particularly manufacturing industries, there tends to be a culture of viewing technology as being the solution to all problems. Whilst it would be naive to argue that this culture does not work (in many instances, it patently does work), it is worth considering the possibility that it is not always entirely appropriate. This may be particularly so when the technological solution is a wholly convergent one which imposes a specific and absolute form of organisation structure on the project regardless of the possibility that the problem may be more appropriately solved by the use of a more divergent solution requiring a less absolute organisation structure. The implications of the imposition of inappropriate solutions can be examined in terms of two forms of energy: synergy and antagonism.

2.1 Energy

Pieters and Young (2000) claim that whenever change occurs, energy is released. This may seem to be contrary to the experience of many project managers who may feel that they have more usually encountered the reverse situation – in order to achieve change, energy must be captured and used to drive a project's operations. They may actually feel that, in a sense, it would be more correct to view the process of achieving change as being one where energy is actually embedded into the project. Such a belief is not completely erroneous insofar as there is a concept of embedded energy that is gaining acceptance in the area of sustainability studies, and it is worthwhile discussing this briefly to reduce the probability of it being confused with the intended meaning of the term within the context of project management as an open system.

In the area of sustainability studies the term 'embedded energy' is taken to refer to all of the energy required to produce, transport and incorporate at its point of use a particular material or component. Consequently, it has far-reaching implications for the operation of any industrial activity, irrespective of whether it would be recognised as being a project. While this is certainly an important area for consideration, particularly in the current climate of concern regarding global warming for example, it is not the perspective to be taken in this chapter. There is no intention to seek some means by which all of the energy involved in the completion of a particular project can be identified and quantified. What is required is more of an emphasis on the energy of the human resource in that this seems to be closer to the intention of Pieters and Young in their discussion of released energy. In the previous chapter the importance of the human resource to project management (such as when considering functional specialism) was introduced, but there is now a need to focus the discussion on the manner in which energy is released when carrying out individual specialisms, and the key words in this regard are suggested as being synergy and antagonism.

When considering energy in terms of a project's human resource, it is important to make fully clear that there is no intention to focus on work-study issues such as how efficient a particular production process may be. Again, this area can be of considerable importance when planning the project activities, but the perspective on 'human' embedded energy to be adopted here is one of its use in either a beneficial (for the success of the project) or a detrimental manner. This emphasis on the human resource is a further development of the assertion in Chapter 1 that information and knowledge are different inasmuch as information becomes knowledge only when intelligence is applied to it. If that intelligence is human, it has the potential to bring along with it a range of values and beliefs that may determine the form of knowledge that the information is turned into. The study of human communication, for example, raises the issue of the encoding and decoding of ideas.

Encoding may be done in many different forms, some of which contain subtle variations that raise possibly significant differences in interpretation (decoding). This can be particularly evident with regard to the use of humour in the management process (humour, believe it or not, can be an important tool in project management and will be dealt with in more detail in Chapter 7). However, when considering the issue of human energy within an organisation structure, if one or more individuals who are part of that structure choose to expend energy in deliberately introducing subtle encoding 'errors' to the information flowing around the structure, the implementation of change can be made significantly more difficult. On this basis alone, the human resource can be considered as possibly the most significant import from a project's external environment with regard to the flow of information around a project organisation structure, and it therefore merits further attention.

2.1.1 External environment resources

The diversity of external resources that may be imported is considerable. Advances in materials technology alone have presented opportunities which did not exist only a few years ago. When advances in other areas are also considered, the extent of resources available in the environment external to any project can be somewhat overwhelming. However, it is equally accurate to assert that not all projects will import all of the resources available to them. The old adage of climbing mountains simply because they are there should not apply when considering the importation of resources into projects. Instead, it is possible to focus on one resource which all projects require: the human resource. It is also possible to simplify this initial consideration by focusing on two aspects of this resource relevant to a project's success or failure: synergy and antagonism.

Synergy in project terms is generally accepted as being the achievement by those involved in a project of the level of commitment where $2 + 2 = 5$, or the sum is greater than the parts. Synergy is of considerable importance to the success of projects as a synergistic workforce can generally be relied upon to be more diligent in its activities than one that is not synergistic. The literature on teambuilding, for example, places great emphasis on the achievement of synergy. Of particular interest in the context of this chapter is the work of D'Herbemont and Cesar (1998) in which the issues of synergy and antagonism are both examined in terms of the energy brought by a player to a project. Before entering into the detail of these states individually, there are two important points to be aware of:

- It is possible for a player to exhibit both synergy and antagonism within a single project, although one will generally exceed the other.
- It is generally regarded as unrealistic to encourage all players involved in a project to become synergistic. D'Herbemont and Cesar suggest that projects will generally contain more people who do not expend much energy than those who do (a ratio of up to $4:1$ is indicated). Pieters and Young (2000) also suggest that where change planning and management on a project are weak, energy tends to be manifested as obstructive and resistive actions by the players. Much energy is expended on the act of speculation, in the absence of reliable information flowing from strong change planning, for example.

Obstruction and resistance can also be viewed as being antagonistic behaviours in that antagonism is generally manifested in terms of not taking the initiative. This may occur in its most obstructive form as being an unwillingness to follow any initiative (change) proposed by others, or in a less obstructive form as being a willingness to follow such initiatives whilst also being unwilling to take initiative oneself. But what is meant by the term 'initiative'? Initiative in this context is defined as being willing to act in favour of the project without being asked, thereby identifying those who do not take the initiative as being antagonistic to a project. Antagonism also has some similarity with the concept introduced in Chapter 1 of mirage projects, in that project-focused antagonism is defined as being the energy an individual is willing to expend in order to make a competing project succeed. Such competing projects may be a variant of the real project which the individual sees as being more important (hence the similarity with mirage projects) or a project which has no link whatsoever with the real project (D'Herbemont & Cesar 1998). The concepts of antagonism and synergy should be considered further when working through Case study 1.

2.2 Throughput

Case study 1: The motive force device problem

Introduction

Case study 1 is intended as a vehicle for the consolidation of some of the system concepts introduced previously and also for the introduction of a number of new concepts, such as maintenance activities. In order to widen the perspective on these issues, an R&D example from the manufacturing industry is used. However, due to issues of confidentiality, all information that may allow any of the players involved to be identified has been given the disinformation treatment. Whilst it is appreciated that this makes the case study somewhat artificial, the outcomes remain relevant to the content of this chapter.

Working as collaborators – Case study 1

The case study examines a project involving International Organisation 1 (IO1) and International Organisation 2 (IO2) working in collaboration to develop what could be referred to as a motive force device for the more classified end of the marine sector. Both organisations are well-established players in their respective markets and this project represented their first significant joint venture. The organisations are of different national origin, with one characteristic of this being that they do not share the same first language. Their collaboration was formalised through an agreement that resulted in project boundaries being established. The project will be referred to here as the MFD 01 project. To the IO1 staff involved in the project it initially seemed to be a typical development project for their company in that they were able to identify a number of typical characteristics. Particularly important characteristics were that the project was focused on a complex product, appeared to be supported by the IO1 organisation, was defined in terms of work/organisation breakdown structures, and programme structures were available.

Each of these characteristics may seem to be straightforward in that you consider yourself to have a clear understanding of the terms used. Take the term 'complex', for example. The MFD 01 project involved bringing together several thousand components, each of which was placed into subcategories of components such as mechanical, electrical/electronic, etc. There are also the issues of performance and quality to consider within the overall problem of complexity. As one example of this, when the MFD operates at its maximum design speed, some of the components have to withstand the equivalent of the weight of a truck bearing on them as they are pressurised. Many project managers would consider it reasonable to view such a product as being both complex and high performance. However, we all tend to use terms rather loosely, and what one person may view as being complex, another may view as being difficult.

This sort of situation tends to drift into conflict as the differences between complex and difficult are sorted out and a common perspective evolves as part of the project culture. Also, everyone involved in project management knows what is meant by the term 'programme', don't they? Possibly not, as we will see when the problems encountered by IO1 on this project are discussed. Those writers who support the move to post-industrial, transformationalist organisation structures may argue that this problem of language will diminish as the emphasis moves increasingly onto the development of knowledge workers and the resulting shift away from hierarchical authority and towards sapiential authority. This is an issue worth being aware of at this point, but a more detailed discussion of it will have to wait until Chapter 4. However, as a sample of the problem's extent, Appendix 1 includes the Association for Project Management's (APM) glossary of project management terms. A scan through this can really start to raise an awareness of the potential problems of language, both in terms of quantity (extent) and meaning.

A final point to consider before progressing onto a more detailed analysis of the project relates to the issue of how an organisation approaches a project. This book addresses factors in the decision that all organisations have to make prior to commencing a new project: how are we going to structure ourselves to deal with this project? Unfortunately, it seems that organisations tend to deal with the problem of answering this question by simply not asking it, or at least not in any meaningful sense. IO1, for example, simply used its standard project team organisation structure apparently without questioning whether this structure was actually the optimum one for the MFD 01 project. The evidence suggests that the question was, in effect, amended to become one of 'can we impose our current standard project team organisation structure on this project?'. IO1 could have avoided problems if it had not asked that particular form of the organisation structure question.

In addition to the above characteristics, the MFD 01 project was defined by agreements between the two collaborators covering a range of issues, including the nature/extent of technical support to be provided, the form of the temporary IO2-IO1 organisation, and an issue that seems particularly dear to the hearts of those involved in aerospace projects: change control standards. Nonetheless, the project hit several problem areas that detracted from its operation and overall success. These problem areas more relevant to this case study are typical of projects in general in that they involve communication, establishing and operating interfaces between systems and processes, the development of a viable project culture, and the perception by each player of the relationship with other players.

Looking at each of these in a little more detail provides a better understanding of the nature of the overall problem from IO1's perspective. The communication problem, for example, can typically be regarded as comprising a number of subproblems (to use systems terminology), such as establishing the formal (as opposed to informal) communication route, the nature of the communication media to be used, and establishing channels between the numerous locations and/or companies involved in the project. The latter subproblem may seem a little unusual for project managers with experience of large construction or infrastructure projects which are typified by the diversity of contractors and subcontractors, extent of specialist suppliers, fluctuating levels of labour on-site during the project duration, and so on. However, it should not be forgotten that IO1 is essentially a manufacturing organisation and as such should not be considered as being automatically in tune with the requirements of a project which it gradually became apparent was somewhat outside its usual type. It is also important to appreciate that problems/subproblems can be inter-related and that the solution for one problem may well have implications for other problems/subproblems. The response by IO1 to the numerous locations/companies subproblem, for example, also has clear implications for the project's culture problem, along with less clear possible implications for some of the other problems.

When responding to the communication problem, IO1 identified several factors within its proposed strategy, one of which was the use of video-conferencing facilities. The conference rooms supplied provided access to video-conferencing facilities, a factor which was considered important in that it allowed face-to-face communication between people who were not geographically co-located, and also reduced the disruption to work caused by the previously regular travel by project players between the two countries in which IO1 and IO2 have their main research facilities. An important point to note here is that there is little evidence that the two organisations were intentionally seeking to create a virtual team environment for the project. Virtual teams will be addressed in further detail in Chapter 4, and at this point it is sufficient to note that structuring for the operation of virtual teams or organisations requires more than simply providing video-conferencing and e-mail facilities. In the case of the MFD 01 project, it seems that IO1 and IO2 were simply trying to increase communication opportunities rather than seeking to achieve a particular project culture or environment based upon the concept of virtual teams.

The issue of culture is one that can be regarded as running throughout the problems faced by IO1, as evidenced by the company's response to the perception of relationships. Within this problem two subproblems were identified. The first can be referred to as essentially being one of seniority within the relationship between IO1 and IO2 as it seemed that some players within the two organisations took the view that they were providing instruction on how to carry out specific tasks, whilst others took the view that they were supplying a service. Such a situation was inevitably a source of conflict between those who saw themselves as instructors and those who saw themselves as suppliers (refer back to the discussion on real vs mirage projects).

The second subproblem was perhaps an extension of the first in that there was some concern about the relationship being interpreted by some as one based on one organisation being the customer of the other, rather than both working as equal partners. This relationship was, apparently, not clearly resolved within the project organisation structure and therefore allowed the possibility of a difference of interpretation. In both cases a relevant factor was doubtless the personal culture of the different players involved. If, for example, one player had always operated as an 'instructor' but was actually required to operate within the MFD 01 project as an 'instructee' (or pupil), it may well resist the apparent loss of status and seek to continue to operate as an instructor. This approach presents the potential for conflict with the actual instructor.

Along with the issues of personal culture, there can also be the matter of national or regional culture to consider. IO1 found that some discomfort was created amongst the players due to apparently insignificant national cultural differences (these were certainly not taken into account when the MFD 01 project was initiated). The IO1 players were in general somewhat uncomfortable with the IO2 players' need to formalise the working relationship through rituals such as shaking hands daily and basing personal relationships on the use of surnames. There was also a perception that the IO2 players suffered from a fixation on accuracy and punctuality, and the final straw for some IO1 players was the perceived absence of any frivolity amongst the IO2 players. So, plenty of clues in there for those who believe in national stereotypes – this is starting to feel like a textual version of the *Through the Keyhole* television show! On a personal note, having worked in the country where IO2 has its HQ, I can vouch for the fact that its citizens do have a sense of humour. Unfortunately for other nationalities, it has to be said that it is a humour that is very much an acquired taste.

The IO1 response to these cultural problems was to target a number of key areas within the overall consideration of customer relations. One of the more significant areas identified was concerned with establishing an understanding of the actual relationship (rather than a mirage relationship) between all players (not simply those working for IO1 or IO2) in the project. A factor in this process was the formalisation of customer/supplier relationships between players. This action produced benefits with regard to both reducing the previously discussed problem of individuals perceiving themselves to be teachers rather than suppliers, and raising the status of customer care activities.

These benefits provide evidence of the generic nature of some responses or solutions to a problem: it is not always possible or even desirable to provide a solution that is specific to one problem. An important consideration raised by this not uncommon situation is that of what conversion process(es) imports are actually required for. This is particularly important when considering the import of human resource energy given its previously discussed tendency to invent its own project.

2.2.1 Import conversion

At this stage of the book it is worth considering imports in terms of two fairly arbitrary categories: human resources and non-human resources. While some reasons that support this division have been discussed previously, it is still arbitrary in the sense that it is subjective – done by reference to a (personal) internal standard. This does not mean that it is any less valid than an objective division; it simply means that you may not agree with it! If not, don't worry as this division is only a temporary one and will be replaced with others as the book (and your expertise) develops.

So, why use this particular division at this point? The primary reason is that of autonomy. Put simply, the human resource is capable of directing and withdrawing its energy with regard to a project, whereas a non-human resource is not. (Ongoing developments in the field of artificial intelligence (AI) are starting to blur this division somewhat, but it is not a significant problem at this point.) As far as the conversion process is concerned this possibility suggests that we may need to put in place an organisation structure which is capable of monitoring, directing and controlling this particular resource in a more open system manner than may be required for a non-human resource. After all, concrete is not going to decide for itself that it is not going to achieve initial set tomorrow morning. However, the batching plant manager may well decide to leave the cement out of the concrete mix tomorrow morning. In either case the end product is essentially the same – an unexpected event that has implications for project change (in other words, a problem) and requires a solution to be implemented as a response.

The conversion process therefore needs to be supported by a number of other activities that the project system must be capable of providing. If the project organisation structure is not designed with an awareness of these activities, it is inevitable that the project (and the project manager) will be faced with problems that could have been designed out prior to commencing work on-site. This sort of problem-avoiding approach is frequently identified as being typical of the Japanese desire to achieve zero defects. The history of it is actually rather more complicated than that (see Chapter 4), but a point worth noting here is that there are believed to be cultural influences within the Japanese approach to organising for projects. These cultural influences appear to drive the Japanese to seek approaches that identify and design out problems (problem avoiding) before commencing production or work on-site, rather than the problem-solving approaches (deal with problems as they arise during production or on-site) which are more typical of European and North American industries. The national culture therefore drives, to some extent, the project culture (as found in the MFD 01 project) and the resultant organisation structure – a further example of a partly open system. Such a system requires two key supporting activities. These can be identified as the maintenance and regulatory activities.

2.2.2 System maintenance activities

Miller and Rice (1970) made an early attempt to differentiate between the types of activity that a system may indulge in, and in so doing identified three types of activity relevant to this discussion. The first type included all the operating activities and these can be considered as being activities that contribute directly towards the ICE processes within a specific system. Consequently, they may differ from system to system and fall within the general discussion of systems to this point.

The second type of activity was identified as being maintenance activities. These are somewhat different to operating activities in that they deal with the procurement and replenishment of the resources that contribute to the operating activities. In the situation where the operating activities may be focused on the production of bread, for example, the maintenance activities would be all those activities dealing with the purchase of flour, yeast, etc. However, by stepping back to the previous consideration of resources as being human and non-human, it can be established that maintenance activities would also deal with activities such as the recruiting, training and (most important with regard to the factor of energy release) motivation of the labour resource. They would also include the purchase and regular overhauling of hardware, machinery and other similar non-human resources. Maintenance activities can therefore be a wide-ranging collection and will usually be specific to a particular project, but there may be instances where a series of similar projects will be initiated by one organisation. In such circumstances it may be possible to replicate all of the maintenance activities across the projects.

Irrespective of any possibilities for replication, the primary function of the maintenance activities that a project manager needs to be aware of (at this point) is that they should ensure the availability of human resource energy at the correct time, in the correct location and of the correct type. This latter point is discussed further below (see 'Driving functional specialism'), but for now it is sufficient to note that any project organisation structure must accommodate the required maintenance activities if the project is not to suffer from human resource energy-related problems. However, maintenance activities are essentially resident in the project's internal environment. Because of this they can function only when supported by a third type of activity: regulatory activities.

2.2.3 System regulatory activities

Regulatory activities were considered by Miller and Rice to be those that:

- relate operating activities to each other;
- relate maintenance activities to operating activities; and
- relate all internal activities of the project or enterprise to its (external) environment.

The regulatory type of activity can therefore be regarded as being largely concerned with information. In the example of a conversion process, the regulatory activities would concern themselves with the gathering of information about the process (import), which would then be compared with relevant performance standards (conversion) before making a decision to stop, adjust or continue with the conversion process (export). In comparison, the maintenance activities would concern themselves with selecting suitable materials (import) for the conversion process, implementing the correct technique of working on the resources (conversion) and directing the completed product onto the next activity within, or outwith, the system (export).

In this example it can be seen that both types of activity can be validly regarded as systems in that imports, conversion and exports can be identified for each. It can also be seen that the relationship between the two may be so close that they are apparently one. However, the project manager needs to ensure that the two types of activity (maintenance and regulatory) are differentiated in order that the project is presented with an optimum environment (in terms of organisation structure) to enable it to function effectively.

In the true manner of a closed loop, Miller and Rice brought both under one umbrella. This was in the form of what they termed 'the managing system'. Such an action should not be regarded as conflicting with the previous assertion that the two activity systems should be permanently differentiated. It is simply the case that each should be aware of the other. An interesting point within Miller and Rice's description of the managing system was the assertion that the system would in effect require to resize itself to the order of operating activities being planned. So, a first-order operating system would require a managing system, but if this operating system was further differentiated to provide a number of second-order operating systems, each of these would also require a managing system.

In short, it seems that the need for maintenance and regulatory activities cannot be avoided, irrespective of the order of operating system to be implemented. Without a managing system the project's operating systems seem doomed to fail through a lack of congruence between maintenance and regulatory activities. This is particularly so the closer a system is to being fully open, and is generally most evident (irrespective of where the system lies on the open–closed continuum) through a failure to achieve the correct output for a given project system.

2.3 Output

The output, or export (there may be more than one), of a project must be readily definable, as the entire effort of the project should be directed at its achievement. As previously discussed, the conversion process will be planned on the basis of identifying the desired output whilst also avoiding, as far as is feasible, any exports that are identified as being undesirable (such as pollution). By applying the systems model it is also possible to look at different orders of conversion process within a project.

The concept of different orders was introduced when considering maintenance activities earlier in this chapter and discussed in terms of scale. Because the systems model of ICE is valid at any scale from a complete project to an individual task within the project, it is possible to consider the issue of output (or export) for any of the conversion processes forming the project. If the conversion process (or processes) is correctly focused, the product can be discharged (exported) either from the project or from one conversion process to the next within the project. There is also the need for open systems to consider the conversion processes that may be placed outside of the project's direct management (in the external environment) but are nonetheless interdependent with those within its internal environment. This was possibly part of the problem experienced by IO1 and IO2 with regard to system and process interfaces on the MFD 01 project (see Case study 1). A key objective for such systems therefore becomes that of discharging the correct product.

2.3.1 Discharging the (correct) product

Quality is a standard criterion with regard to the discharging of products from a system. Chapter 1 introduced the concept of boundary controls and how these could be used to ensure that only a suitable quality of import was allowed into a system, and an export was discharged (exported) only when it was also of a suitable quality. The boundary control concept can also be linked with that of regulatory activities, in that these have been previously stated to be concerned with information. From that point it is only a relatively minor hop to the point where boundary controls can be regarded as a system themselves – they gather data, compare it to a predetermined standard, and then determine an action on the basis of that comparison.

The link between regulatory activities and boundary controls is further strengthened in situations where a project has multiple interfaces with its external environment (regulatory activities have been characterised as relating project internal operations to the external environment). Such interfaces will typically be concerned with importing resources from the external environment, such as is the case with both materials and the human resource. These import interfaces will be defined in terms of information concerning the particular conversion process for which the resources are required, and can therefore be regarded as allowing for the discharge of a resource from outside the project. The project therefore sets the standard against which the (external) product's quality is judged and a regulatory activity will be involved in making the external environment aware of that standard.

Other, probably fewer, interfaces will be concerned with the export of either partially converted products (that may require further, specialist processing outside the project environment) or of some undesired and incidental export such as waste materials or pollution. In this case a regulatory activity will also be involved, but the emphasis now becomes one of what quality the external environment will deem acceptable before it will allow the project's export to become its import. Nonetheless, the key consideration is that of information – the correct information needs to flow through all interdependent operations. If the project organisation structure allows and indeed encourages this, energy will be released as the human resource is able to go about its various activities. When the project organisation structure does not allow this to happen, energy is still released but it is wasted in trying to overcome the blockage, leaving less energy for the completion of the required activities elsewhere. The emphasis therefore should be on the facilitation of (correct) product discharge.

An important aspect of the facilitation process is the realisation that projects happen in real time (D'Herbemont & Cesar 1998) and this makes those projects operating as open systems susceptible to changes in the external environment. Consequently, it is not always possible to find the planned-for resources within that environment and the project may have to consider allowing the import (discharge from the external environment) of one or more resources that are in some way different from that intended. It may be tempting to view such a situation as a failure of the project planning function – someone must be to blame and should be punished. However, there is only so far that the planning function can go in achieving a problem-avoidance approach to the management of a project in real time, and at some point reality will inevitably jump in and seem to mess things up.

It is at this point that the project needs to allow for human creativity to begin releasing a little energy, and it has been suggested that projects should be managed by placing the optimisation of the release of energy by the players first, and the focusing on technique (such as project planning) second throughout the duration of the project (D'Herbemont & Cesar 1998). One aspect of such an approach is to be aware that a manager's perception of what he or she wants to do is usually considerably different from the perceptions of the other players regarding what the manager wants to do. This situation raises the need to consider events and their management.

2.4 Event management

Project managers can tend to take an approach that concentrates on the so-called big picture. This approach usually involves delegation of the detail and minutiae to others lower down the project organisation structure or hierarchy, and has led to the perception of project managers by some practitioners as typically being 'charismatic fire-fighters'. Indeed, the ability to delegate is often cited as an important management skill and there is no intention here to suggest that project managers should not delegate. After all, this has to be one of the benefits of functional specialisation – if there is someone in the project organisation who is more skilled in the use of planning software, it makes sense to delegate that task to them. However, it is important that managers do not delegate involvement as well as tasks down the hierarchy. This is particularly so when considering the concept of events.

In the context of project management generally, events are interpreted in terms of planning procedures, such as can be found in the terminology for critical path methods. In this terminology an event is something that occurs at the beginning (the start event) and conclusion (the finish event) of an activity (Pilcher 1992). It may also be taken to be an exceptional fact concerning all players in the project, such as a fatal accident in the workplace. There is, however, a further interpretation of events as being facts of limited scope, in that they affect only a few players, each of whom may also link them to a specific date (D'Herbemont & Cesar 1998). An example of such an event may be the introduction of a new coffee machine for the management offices. The machine's arrival is a fact (it is either on-site or not) that is of limited scope (only for the use of management staff) and can be linked by the affected players with a specific date, if there is some particular reason for doing this (lack of coffee!). A more rigorous reason can be suggested in that the machine's procurement will have involved both maintenance and regulatory activities, and as stated previously, regulatory activities emphasise project information. For example, the machine may be expected to arrive on a particular date, or perhaps even at a specific time, but if it has not arrived by that date/time, a response activity should be implemented (contact the supplier, perhaps). In the case of the machine's non-arrival the expected event is replaced by an unexpected event. Finally, when the machine is actually onsite, its presence ceases to be an event and becomes a simple fact.

The machine's arrival on-site is hardly a major event for the project, so should it be of interest to the project manager? Rather than consider the importance or scale of the event, the project manager would do better to consider the significance of the event in terms of it being either a freestanding or a chained one. If the event is freestanding (is not linked to other events happening previously, at the same time, or in the future), then an awareness of it could be useful. If it is a chained event (linked to one or more other events), then it becomes less relevant in that the project manager need only be aware of some, rather than every, event in a chain.

The reason for developing this awareness of events is that they tend to cause tension amongst those concerned with them (D'Herbemont & Cesar 1998). Tension within the human resource can act either as a positive force (in that it spurs creativity) or as a negative force (in that it diminishes creativity). The project manager needs to be aware of events so as to be able to minimise negative tension and maximise positive tension. In this way energy is released in a manner enhancing the completion of project-relevant tasks (emphasis on maintenance activities) rather than one focusing on the removing of blockages (emphasis on regulatory activities). This can be regarded as optimum event management.

The process of event management can be structured around seeking to identify events under four headings:

- Events directly associated with a project and predictable prior to project commencement. Generally the simplest events to manage due to the time available to plan actions (if the project team manages actually to predict them).
- Events directly associated but not predictable. Generally the most difficult to manage because of the lack of time to plan responsive actions and the fact that they are usually of a negative nature. This event type is similar to Dainty and Moore's (2001) concept of the unexpected change event (UCE).
- Predictable events not directly associated but having a strong influence on project players' interpretation of the project.
- Events which are not predictable and not directly associated.
                    (Adapted from D'Herbemont & Cesar 1998.)

Throughout each of these there is a common thread: information. If the project organisation structure does not allow for the free flowing of relevant information, it will be difficult for the project manager to carry out successful event management, and the project's human resource energy may become increasingly drained through dealing with negative rather than positive tension. This suggests a link between information and energy, and the nature of that link can be considered in terms of entropy and communication noise.

2.5 Entropy and the issue of noise

At this point an item is included to comfort those readers who are of an engineering background: project management and the second law of thermodynamics! It's not every day that you come across these two areas being linked, so it may prove to be a novel experience for some. A good starting point would therefore seem to be a brief description of the second law of thermodynamics. This states that the universe is winding down and becoming increasingly disordered (which possibly also explains why junk mail companies are tending to send out junk addressed to 'the occupier' rather than using your actual name? Sorry, a digression – this happens from time to time). Useful forms of energy are slowly turning into useless forms (from the perspective of doing work), a situation identified by a concept called entropy. Any system that possesses order in its operation, such as human beings, is said to have low entropy. Unfortunately, the universe doesn't seem to like low-entropy systems and is constantly trying to convert them to high-entropy systems. If it is accepted that both closed and open systems have order in their operations, they can both also be classed as low-entropy (or at least lower-entropy) systems and are therefore targets for the universe's desire to create high-entropy systems. Perhaps unfortunately, closed systems by definition cannot be aware of this situation, but even if they could, they would not be able to do anything about it.

Open systems, on the other hand, have the potential to be aware that they are at risk of being converted. This potential arises through the possibility of them being, to varying extents, sentient or self-aware (this point is covered in more detail later in this chapter and will be returned to in subsequent chapters) – they know they exist and wish to continue existing. However, they can continue to exist only if they can manage to convince the universe that they are actually high-entropy systems. It has been suggested that we humans do this by 'adding in' disorder to everything we do. In effect, we create high-entropy systems around us to disguise the fact that we can survive only as low-entropy systems – we waste energy. Humans are not alone in doing this, as all living things perform basically the same con-trick on the universe (Ward 1999), but we are probably better at it than most because of the technologies that we surround ourselves with in our daily lives. In other words, we are a more open system than most on this planet.

Applying the second law to the organisation of projects suggests that any attempt to provide order (in the form of a low-entropy system) so that production processes can operate without interruption, for example, is immediately faced with the universe's dislike of low-entropy systems. A significant consequence of this is that organisation structures are always seeking to balance the need for order to support their production or project processes, and the subsequent natural tendency for such structures to collapse into disorder. Project management is perhaps particularly problematic in this regard and can therefore be viewed as operating on the knife-edge between order and disorder. This does not mean that any attempt to project manage is ultimately doomed to failure – the whole range of technologies with which we are surrounded on a daily basis proves that this is not the case.

Nonetheless, there is the ever-present spectre of failure, and whole books have been written on the many spectacular failures that have occurred during the human race's recorded history. One small example is the case of a Roman engineer who set off to tunnel through a hill. Being an efficient sort of fellow he decided that the minimum project duration would be achieved by setting to work two teams of tunnellers, with each team working from opposite sides of the hill at the same time. With hindsight it is easy to see the potential for disaster, and the project did indeed meet with the almost inevitable result that the tunnels failed to meet in the middle. Fortunately, the Channel Tunnel project fared rather better in this respect, so obviously some progress has been made.

There can be benefits for the function of project management from the constant struggle between order and disorder. Moore and Hague (1999) have suggested that the most creative forms of management can be found being practised in environments where the rate of change within a project or process is right on the boundary of being classed as stable, or predictable, and able to become unpredictable with no discernible warning at all. The implications of such a situation for the design of an organisation, particularly with regard to exercising the control function, are significant and will be examined in more detail later. However, it is safe to assume at this stage that such environments require the ultimate form of knife-edge organisation in order to mediate the extreme demands being placed upon projects operating within them. This introduction of the mediation role can be developed further to arrive at a suggestion that many with a traditional project management background may find difficult to accept: organisation structures are not intended to control a project.

The emphasis within this book is on the designing of appropriate project organisation structures to enable mediation between the various demands placed on the project. The emphasis is not upon controlling the project through the organisation structure, as this function is arguably more correctly viewed as being reliant upon the people within the structure rather than upon the structure itself. This is in accordance with an emerging view of organisation design suggesting that, while project managers need to understand a small number of key theories about the operation of organisations, they also need to resist the desire of rational thinking to use this understanding in controlling and predicting organisational events. In fact, this emerging view goes so far as to assert that the reality of organisations will not conform to any logical or systematic patterns of thought. This comes about because organisations are more accurately viewed as being relatively small patterns of energy operating within the larger pattern of the environment (Banner & Gagne 1995). This is not far removed from the earlier consideration of both the release of energy by the human resource and the effects of the second law of thermodynamics on open systems.

The pattern of energy perspective on organisations is supported by those who can be categorised as transformationalists – a rather uncomfortable title perhaps suggestive of obscure religious cults. Irrespective of this, transformationalists argue for a different approach to thinking about organisations, and this has implications for the designing of organisation structures. The traditional (sometimes referred to as Newtonian, but more usually as transactional) approach to organisation structure design is to concentrate on the various forms of structure available, such as the hierarchy of an organisation, or its complexity. However, this approach can cause difficulties in a number of areas due to the relatively inflexible nature of the resulting structures.

One particular difficulty is that of recognising problems during a project. The structure must intervene in some manner between the problem and the project if the project is not to be adversely affected. However, if the problem is not accurately recognised, there is the probability of applying an incorrect solution to the problem. Problem recognition is largely related to how information is gathered by an organisation – if unclear information is collected, or clear information is gathered but then is communicated badly, problem recognition will be adversely affected. Awareness of this situation further reduces the effectiveness of problem recognition in that the organisation may fall into what is referred to as perceptual defence mode – the organisation's tendency to deny the existence of what it perceives as threatening situations (a form of incorrect event interpretation). The project team then releases the majority of its energy in either dealing with virtual projects or overcoming negative tension. Once the project team slips into perceptual defence mode, the project will rapidly descend into the dreaded state of disorder. In this state it is unable to respond to problems and threats because it does not acknowledge them as existing. In essence, much energy is wasted on dealing with problems that have arisen as a result of 'noise' within the project's communication flows.

The transformational approach seeks to bring about the opposite situation where all the players are able to exercise their innate creativity in the solving of problems within the context of a fluid and creative entity. Players are focused upon the vision or goals of their organisation rather than upon competing with others within that organisation, or seeking to complete mirage projects, and thereby achieve effective organisational functioning. Perhaps more importantly, it is claimed that this is achieved with less emphasis on formal structure, rule, policies and so on than is the case in the traditional approach. An important task within the transformational approach is to ensure good communication with regard to a key aspect of projects and entropy, and this relates to the concept of a project life-cycle.

The natural progression of an entity's life-cycle involves the depletion of energy available until the point at which the entity dies – it is no longer capable of useful work. Applying this model to a project suggests that during the early stages of the project the level of energy available (certainly on projects fitting the S curve representation) increases – more resources are brought into the project and it grows rapidly. However, as the project enters its later stages, resources start to be removed from it (people and equipment, for example, are moved to other projects) so that its level of available energy diminishes until the point at which maximum entropy is achieved (the project is completed and no more useful work is carried out in the project environment). It can therefore be argued that projects naturally increase in entropy, and that a successful project is in fact one that has achieved the intended level (of entropy) as represented by completing the planned change of inputs into the desired output(s). It is therefore quite possible to manage a project's growth through the control of information (and the resulting expenditure of energy) concerning the project's requirements.

One of these requirements is that the project must end at a specific point and this must be clearly communicated to all players so as to manage the 'end' event as being a positive one, thereby seeking to overcome the tendency of some players to resist it by searching for ways to extend the project. This typically involves the development of virtual project activities, for example – a form of extension that can be classed as evidence of the human resource seeking to import excess energy into the project environment.

2.5.1 Importing of excess energy

The concept of entropy needs to be balanced by a consideration of its opposite state: negative entropy (sometimes referred to as negentropy). This concept is almost one of those examples of two wrongs making a right, in that negative entropy is actually a positive attempt to resist disorder and impose order on a system that is being pushed by the universe to become high-entropy.

One problem with negative entropy is the determination of how much is required by a project, or by any other entity. In other words, how much excess energy will the project require so as to be sure that it goes through its intended life-cycle and achieves a high-entropy state at the planned rate of change rather than at an accelerated or decelerated rate due to unplanned or unforeseen events? This problem is actually rather more complex than it may at first seem due to the need for the majority of projects to also contribute to the continuation of their parent organisation(s). As mentioned previously, projects have a finite lifespan, whereas parent organisations naturally seek to achieve an infinite lifespan, and one contribution to this may be the successful completion of projects. Such projects contribute to an extended parent organisation lifespan through their generation of reserves that enable the parent organisation to stall or delay the entropic process. However, they can do this only if they are successful as projects in themselves, and to achieve that state they must consider the extent of excess energy they are willing to risk importing to their internal environment.

The problem has similarities with that faced by project planners when they seek to maximise the use of resources in individual projects: how much of a safety margin do they wish/need to allow when calculating the duration of activities? Techniques such as PERT (probability evaluation review technique) allow planners an opportunity to take into account an element of risk through the determination of a duration which has an acceptable probability of being achieved. Within a complete project, such planning techniques allow the planner to experiment with the number of critical activities and decide upon the extent of spare capacity (in terms of float on non-critical activities) with which they feel comfortable. While it is obviously arguable that each of those activity durations – and thereby the project duration – is dependent for its achievement upon the energy (in terms of resources) imported, it is unlikely that project managers view them as also being a representation of excess negative entropy. However, they almost certainly will, at some point, view them in terms of cost, thereby giving the opportunity to consider the problem of how much excess energy to import from a potentially more familiar perspective: how much surplus capacity can the project afford?

2.5.2 Effects of surplus capacity

Surplus capacity can be regarded as a buffer against unplanned or unforeseen events and in that context can be regarded as an essential weapon in the project manager's armoury. Unfortunately, it also costs money. As a general rule of thumb (heuristic), the more surplus capacity a project carries, the less likely it is to be successful (make a profit) and thereby contribute to its parent organisation's ability to stall the effect of entropy.

There is, however, an optimum level of surplus capacity that a project should carry in order to maximise its probability of achieving success. Unfortunately, this level seems to be decided in many cases on the basis of how little money the parent organisation is willing to allow the project to spend (or, in the perception of some, waste) on purchasing surplus capacity. In this sense, an important effect of the surplus capacity decision is to act as a potential constraint on the project organisation structure. Questions may be asked with regard to whether particular functions are actually required, or whether two or more functions could be carried out by the same human resource(s). If they can be combined, or the decision is made to force their combination, the project organisation structure may be slimmed down as functions disappear. In such situations, with the inevitable concerns regarding functional efficiency, the feedback process should be emphasised within the project structure so as to identify unplanned events (such as may result from an individual's inability to carry out two or more functions effectively) at the earliest possible time.

2.6 Feedback

Feedback can, as with most aspects of organisations, become a complex function to implement. Thus far the emphasis has been on ensuring feedback within the project's internal environment, but it is also worth briefly covering the issue of feedback from a project's external environment. Pieters and Young (2000) suggest that external customers are a good source of feedback for an organisation, particularly with regard to assessing the impact of decisions made by the organisation and the responses it makes to decisions implemented by its customers. This is doubtless a useful perspective for parent organisations, but may be argued by some to be less useful for projects, particularly for those projects that would traditionally be regarded as significantly independent of parent organisations.

While there is merit in being cautious with regard to accepting changes in approach, there is also merit in practising innovation and creativity. This is particularly so in industries traditionally regarded as being constrained by established culture in the form of custom and practice. When custom and practice in an industry actively discourage innovation and creativity, it is arguable that it is gradually accreting ways of working which will diminish its ability to respond to sudden and unexpected change. In the spirit of challenging the suggestion that feedback from customers is of little value to a parent organisation's projects, a change of emphasis is suggested by regarding the suggestion in the context of an 'external' customer.

The previous discussion of internal and external project environments involved a consideration of boundaries. Taking that discussion further, in considering the idea of boundaries between systems as being a possible differentiation between a parent organisation and its customers, should not be too problematic. After all, a parent organisation and its customer organisations can be clearly differentiated on the basis of territory if nothing else. However, the concept of boundaries can be developed still further and, when combined with the language of quality assurance (QA), can become a means of differentiating between subsystems within a project environment as being each other's customers. This concept of the internal customer can furthermore be regarded as being a development of the interdependence feature also previously introduced.

In this manner, individual subsystems feed back to their supplier subsystem(s). This type of feedback may be more readily accepted by some as being relevant to ongoing, long-term production processes (where there is time to respond to feedback by making changes to the output) rather than the relatively short-term processes (where feedback may arrive only after the process has been fully completed) involved in typical projects, in which case it may be regarded as being part of the management of the *status quo*. Customer feedback may therefore be linked in the mindset of some individuals to functional, rather than project, management. However, such a viewpoint can be argued to ignore two considerations of possible benefit to project-based organisations:

- Feedback may well come too late for the processes within a current project, but its value for improving the relevant processes in future projects should not be ignored. Carrying feedback forward to future projects can be seen as being a step towards the development of a learning organisation. This point will be returned to in Chapter 7.
- The implications of trying to impose a linear model (functional management) on potentially non-linear project systems (project management).

The latter point pushes the discussion close to the issue of chaos, but rather than getting deeply involved in that at this stage, a gentler introduction will be provided through a consideration of how a project's internal subsystems may be linked to the need for feedback.

2.6.1 Comparing internal subsystems

Internal subsystems are traditionally viewed as being planned so that they can be controlled through comparison: an ideal is planned and then the actual process as implemented is compared to the ideal, thereby allowing any deviation from it to be deemed a valid reason to intervene, thus returning the process to the planned ideal. But what if the planning process itself is in error and the resultant ideal is in effect another example of a mirage? Considerable effort goes into the planning of a project, largely so that the perception of uncertainty can be reduced to acceptable levels for both the client and project team(s). However, that effort is meted out on the basis of a number of assumptions about the project, and perhaps the most important one to consider at this point is the assumption by all concerned of linearity within the project.

Parker and Stacey (1994) suggest that researchers in many fields have previously used the concept of linearity to explain the functioning of systems that they knew to actually be non-linear. The reason for their use of linearity-based explanations was simply that they also knew non-linear relationships to be fiendishly difficult to describe. So long as all involved accepted that linear expressions could be used as approximations of non-linear relationships, they believed there to be no problem in assuming the two types of relationship to be similar. Unfortunately, there are some important differences between the two:

- Cause and effect – in a linear relationship a given action has only one outcome, but in a non-linear relationship one action can have many different outcomes.
- Solution equations – linear equations have only one solution. Non-linear equations have more than one solution and there is no general method which can be used to solve the majority of them.
- Additive property – linear relationships or systems are the sum of their parts and each part can be individually studied and described so as to construct a description of the whole (the basis of typical systems analysis techniques). Non-linear systems are more than the sum of their parts in that they are synergistic and such systems cannot be described by studying individual parts of them. Instead, there is a need to take an approach that deals with the patterns of behaviour possible for the whole system.

(Summarised from Parker, Stacey, 1994.)

On this basis, the traditional reductionist approach to analysis and planning of projects will be viable only if the projects in question are truly based on linear relationships between their internal subsystems and also any relevant external systems/subsystems. It is therefore possible that even apparently sophisticated techniques such as PERT may have no relevance to many of the projects for which they are currently being used. This suggestion is made on the basis that such techniques place considerable emphasis on the use of statistical analysis (PERT, for example, relies on the use of standard deviation and variance for each activity identified). With such analysis there is an identified level of error that is deemed acceptable by all involved and can therefore be ignored.

However, it is becoming increasingly accepted that many systems are not actually linear. For these newly identified non-linear systems, the level of error that was deemed safe to ignore when they were considered to be linear now becomes dangerous. This is due to such systems being extremely sensitive to the start conditions at the point when the system is implemented. Any error or noise in those start conditions can be multiplied up as the system progresses through its subsystems during the unfolding of a project. This occurs due to a potentially large number of outcomes that are possible from any input to each subsystem. It is quite possible that the link between cause and effect, which is crucial within assumed linear relationships, disappears completely within the complex range of interactions that are possible as the non-linear project system unfolds.

Non-linear systems can then be argued to be highly sensitive to start conditions and this needs to be accepted when considering the issue of feedback. In essence, such systems cannot be successfully planned in the traditional sense of the term, in that their performance cannot be driven to achieve some predetermined standard. The emphasis moves away from planning the system so as to provide a framework for its control and towards contributing to the evolution of processes of self-organisation within the system (Parker & Stacey, 1994). Such a scenario has one important feature: the link that is traditionally implied within the process of feedback (outcomes can be predicted on the basis of actions taken and inputs supplied) is broken and nobody can claim to be in control of the project system. Quite a scary thought for those with a more traditional (contingency) perspective on project management, but in actual fact it may not be as scary as it first seems due to the concept of steady state.

2.7 Steady state

This concept is one of four activity types that characterise organisations (Handy 1999). The other three are innovation, crisis, and policy. Steady state is of interest at this point due to its emphasis on those activities that are capable of being programmed in one way or another. They are in essence routine, rather than non-routine, activities. The importance of steady state to the analysis and structuring of organisations lies in Handy's assertion that this type of activity will usually account for 80% of an organisation's personnel. This represents a significant release of energy in dealing with everyday, routine activities. Typical examples of such activities are secretarial, accounting, office services generally, the majority of the production function and most of the sales function (if any). Marketing activity, in contrast would not be deemed to be steady state, as the majority of this function would be classed as innovation activity.

An obvious consideration when considering the suggested proportions for each of the four types of activity is that Handy's emphasis is suggested as being on long-term parent organisations, rather than on relatively short-term project organisations. Nonetheless, the concept is generally transferable to project organisations so long as the user is aware that a particular project organisation may be lacking in some functions, such as marketing for example. A less obvious consideration is that the steady state can be a powerful attractor for those organisations who focus significant effort onto trying to bring all of their activities into this category. Unfortunately, this is defeating the object of having identified four activity types to begin with. Handy suggests that the management of steady state activities should be focused on the application of rules, procedures, regulations and other formal control methods. Innovation activity on the other hand can be allowed to operate in a more informal, task-oriented (rather than method-driven perhaps?) manner. This type of activity would simply be choked by an approach suitable for steady state activity, and can therefore be classed as similar to the previously introduced concepts of creativity and energy release.

It is a small step in terms of reasoning (but a large step in terms of culture for some organisations) to envisage the steady state activity as being in tune with the concept of linear relationships discussed previously. Such activity will generally be typified by cause and effect links between inputs and outputs; do 'x' in 'y' manner, and you will achieve 'z'. Innovation activity however can be regarded as being more in tune with the concept of non-linear relationships. There may be little or no identifiable link between cause and effect with regard to inputs and outputs. This then suggests that there is scope within a project organisation structure for those who seek to control through the operation of cause and effect links, and that they can do this by identifying and dealing with all those activities that can be truly regarded as being steady state. Just don't let them get involved with any innovation activities!

2.7.1 Resource clash

The prospect of resource clash occurring represents a decision-making opportunity, and should be regarded as being inevitable in a project management environment that is bounded in terms of finite resources. Within any production activity there will, at some point, be the need to consider the optimum use of resources. The resource clash concept has previously been considered in the context of nonhuman resources, particularly regarding the sustainable use of indigenous resources in developing countries (Moore & Ahmed 1996). If, however, it is accepted that all resources represent energy and that their individual value is to some extent determined by their ability to turn that energy into useful work (refer back to the discussion on the releasing of energy in completing a project), then a potential problem arises if a particular resource is able to contribute to the project in more than one way. In such circumstances it may be feasible to consider the existence of resource clash; a decision has to be made regarding which of the possible contributions that a particular resource is able to make represents the optimum use of that resource.

It seems reasonable to assert that the resource clash problem is not one that a project manager will be faced with frequently. This is because of two factors:

- increasing functional specialism; and
- human resources tending to possess primary and secondary roles (functions).

The concept of functional specialism has been discussed previously, and will be returned to elsewhere in the book, but at this point it is worth stating simply that increasing functional specialisation by the human resource narrows the contribution that any specialism is able to make to a project. The result is that there is less potential, or actual, overlap between specialisms; the boundary of one specialism is clearly defined and does not overlap with the boundaries of other specialisms. Primary roles for individuals are therefore clearly established and these roles should not clash. However, it is possible that changes in the external environment during the lifetime of a project result in a situation where a particular human resource required by the project cannot be imported. At the time of writing, for example, the UK construction industry is claiming to be in the midst of a severe skills shortage in specialisms such as construction management. This compares with a surplus within the specialism of architecture, and some universities are tapping into the situation by offering hybrid courses such as Architecture with Construction Management. The resulting human resource could be expected to have a primary role of architect, but may also be capable of undertaking the specialism of construction management as a secondary role. In other words, the project manager will be faced at times with being unable to import the required primary role, and will therefore need to be sufficiently flexible to consider importing alternative resources that are able to carry out a relevant secondary role.

The concept of primary and secondary roles is more commonly considered in terms of developing teams. Handy (1999) suggests that individuals will generally operate in terms of a number of predetermined roles when asked to work in a team and that most are capable of changing between two or more roles. They will however, usually prefer to operate in terms of what they see as being their primary role unless there is a surplus of that particular role within a team's membership. In such circumstances, they may be willing to change to one of their secondary roles. It is important to note that primary and secondary roles in terms of team development will not generally equate to primary and secondary roles in terms of functional specialism. Failure to appreciate this can cause severe problems for the project, and it can therefore be useful to consider analysis of the human resource on the basis of differentiation prior to making decisions concerning imports to the project.

2.8 Differentiation – further discussion of Case study 1

Having covered the issue of functional specialisms from several directions, it should now be useful to re-examine Case study 1 so as to gauge some of the results of differentiation for the internal subsystems of a project. IO1 are not unusual in that they seek to take what could be argued to be an essentially reductionist approach to organising for projects, and this approach is covered in several of their in-house publications (confidentiality precludes referencing these). Such an approach is typical of the majority of medium to large organisations who seek to control through the use of rules and procedures. However, it is not typical for all organisations and this issue of approach may be regarded as one issue of culture that is particularly important in multinational projects such as MFD 01 (see also the discussion of multinational organisations in Chapter 3). In the case of IO1, these rules and procedures are encapsulated within a number of frameworks that they claim take a systemic approach to project management. However, they also give the impression that IO1 view the project management process as being linear rather than non-linear. The result is a framework that encourages differentiation through the identification of multiple responsibilities, and does so in a way that suggests it was inevitable that the MFD 01 project would identify communication as a problem.

The starting point for the document is to link the product development (PD) and project management (PM) processes into what is referred to as an integrated process. Both processes share a number of stages (from pre-tender/new product development to aftermarket/in service management) along with several gates, each of which lists criteria to be achieved before it can be exited. Each stage is described in terms of key features, outputs and responsibilities. To support all of this there is also a specific terminology developed. For example, one criterion refers to 'differentiators' with regard to output.

Perhaps the most striking aspect of this framework is the extent of information gathering involved. Pieters and Young (2000) refer to the effect of information gathering on organisations and identify the following repercussions:

- greater use of internal specialists and external consultants;
- added data-collecting responsibilities; and
- added resources to process and interpret gathered data.

It is suggested that without these additions an organisation may gather data but does not develop intelligence, and its ability to deal with the increased uncertainty (represented by gathered but unprocessed data) is reduced. It is to IO1's credit that their framework does address this latter point by adding the requirement to identify lessons learned and feed them back into the organisation, so there is some evidence of attempting to develop intelligence on the basis of the data gathered. However, the gathering and processing activities add to differentiation within the organisation, and also suggest a belief that the outcomes of the project management and other processes can be controlled and affected by their inputs. In other words, the framework can be argued to impose linearity on projects which may in fact be largely non-linear. This suggestion can also be taken to indicate an approach that is more akin to configuration management than pure project management. Configuration management concerns itself with change control executed in such a way as to ensure that a whole system retains consistency between all of its components when one or more components are subject to change (Field & Keller 1998). Whilst IO1 are not operating within the aero-space sector, there is evidence that the aero-space industry as a whole values configuration management highly (Chiavola 2000; Sorensen 2000) and in this respect IO1 would not be considered unusual in the approach taken to managing their projects, regardless of any impact it may have upon levels of functional specialism.

2.8.1 Driving functional specialism

The positive reasons in support of increasing functional specialism have previously been discussed and it has been accepted as inevitable that, in the foreseeable future, this aspect of industrial evolution will continue to increase. Of itself this is not perhaps a significant problem for projects or society as a whole. This suggestion arises from the perspective that functional specialism is one device in the attempts by humans to convince the universe that we are not low-entropy entities. Functional specialism can be argued to do this by increasing negative entropy through imposing increasing levels of order on the wider environment in the form of exploiting gaps in the skills matrix available to projects and organisations. In this perspective, it simply identifies opportunities (gaps) for imposing more order (task differentiation) by releasing more energy (greater number of specialisms). There is also the argument that organisations repond to change in their external environments by adding complexity. This is readily achieved by adding increased numbers of specialisms. However, between each specialism lies a boundary, and boundaries can be a cause of noise in communicating production or project information. Think about that one for a while before moving onto the resultant potential problem of fragmentation.

2.8.2 Fragmentation of resources

Fragmentation can be considered to occur when the interfaces between interdependent resources (particularly human resources) are disrupted. Traditionally, this does not seem to have been a significant problem in that there were relatively few interfaces between the few interdependent resources required by the majority of projects in the past. In this sense, such projects could be regarded as being low-entropy in that they released a comparatively small amount of energy. However, this should not be taken as suggesting that fragmentation was impossible in such traditional projects; wherever there are interfaces, fragmentation is possible. It is simply a case of being unable to identify frequent examples of this unless poor design skills are included, as this seems to have been a common factor in project failures. As was shown in Chapter 1, there were many good examples of project management in historic projects. Likewise, numerous examples of good engineering can be cited, but if design skills are taken as lying somewhere between project management and engineering, then various failures can be attributed to them. Failure may well have been due to an innovative designer not having available to them the project management and/or engineering resources to increase the probability of their design being successful. It has been stated previously that many building projects of the mediaeval period, for example, were essentially exercises in prototype testing to destruction. This was largely because the deterministic skills of engineering were simply not available to designers. However, as time passed and more specialisms were introduced into projects, the extent of interfacing also increased along with the possibilities for fragmentation. At times it seems as if the second law of thermodynamics is always demanding a payback for any advances that are made by a society's process of production.

The extent of specialisation in modern projects is perhaps inevitable given the rate at which the external environment is changing, but specialisation can only be made to work if the resources do not also become fragmented. This may occur as a result of noise within the project's communication networks, and so the project organisation structure must not encourage noise of this (or any other) form. An important step in this regard is to encourage the recognition of what is not known.

2.8.3 Recognising the unknown

Traditional, or transactional, project structures were developed in response to the environment around them. Part of their development was the attribution of responsibility. As soon as a hierarchy is produced, people seem to start thinking in terms of what is and (more importantly for some) what is not their responsibility. Along with responsibility comes the worry about being punished for failing to meet it. There is also the mentality that you do not question an instruction that comes from higher up in the hierarchy; it may seem like a nonsensical instruction to you, but that is seldom seen as a valid reason for ignoring it. In a transformational organisation structure, this would not be the case due to the emphasis placed on knowledge. If an individual knows that an instruction is nonsense, and can demonstrate why it is nonsense, then their knowledge gives them the authority to do something (other than simply carrying out the instruction regardless) about the situation. In this sense, those within the structure are pushed to recognise where their knowledge boundaries lie, rather than merely carrying out their function as defined by a job description for their position in the hierarchy. This willingness for a project's human resource to recognise subjects about which they know nothing (in the words of Manuel the waiter) is probably an effective way of reducing the extent and impact of fragmentation. Unfortunately, it seems a difficult feature to imbue to traditionally structured project organisations as it impacts on the very culture of the project team itself, and is generally resisted.

2.9 Equifinality

Put simply, the concept of equifinality asserts that even though different starting conditions may be used, and different pathways are taken, it is still possible for two or more projects (or organisations/entities) to finish with the same result. It is also possible for one project to be planned to be completed using two or more different approaches, and still produce the same result. In short, there is seldom only one way to achieve the required outcome(s). The 'one way' scenario tends to result from situations where the outcomes are defined so narrowly that there is no scope for variety of approach to achieving them. It also results from dealing with projects that are highly linear; the scope for alternative approaches can be highly constrained, if not completely eliminated.

While human creativity can be a powerful force, human obsession can also be a powerful constraint. One example of this can be found in the different approaches to specification writing. At one extreme lies the totally performance-oriented approach wherein the sole consideration is to achieve a particular outcome, or group of outcomes, without any particular methods or materials being imposed. The desired outcomes may be expressed quite simply, as in the case of desiring to be able to shelter somewhere warm and dry when the weather becomes cold and wet. Such outcomes may be achieved in a variety of ways and using a variety of materials; everything from a tent to a 747 would provide the required results.

The expression of desired outcomes in such a simple manner could be expected to result in an open system for their achievement; the project is relatively free to interact with its external environment in a wide range of ways. However, this introduces a few problems, and the two main ones as far as project structures are concerned can be suggested as being linked to the issue of pathways.

2.9.1 Pathways

The issue of pathways within project structures is not something that traditionally structured projects have previously given much in the way of explicit consideration to. This is simply due to the fact that they have rarely felt the need to do so, particularly in the case of projects driven by a traditional form of specification in which materials (type and quality) and methods are highly specified. Such projects can be regarded as being at the closed end of the system continuum and they seek to control both their internal environment and all interactions with their external environment. A significant implication of closed system projects is that the organisation structure is largely determined by the specification, and in this regard it could be argued that such projects are essentially deterministic in nature. The result being that there are few, if any, alternative pathways available to the project team. Projects of this nature tend to be large and complex, if only because highly detailed specifications take time and money to produce and can be argued to represent an unreasonable overhead on smaller projects. The traditional approach goes something along the lines of there being little point in spending more money on the specification than it would be expected to save in terms of reducing costs attributable to factors such as redoing incorrectly executed work.

When dealing with open-system projects that are not so constrained by detailed specifications there is a need to make decisions concerning the pathway to be taken through the myriad possibilities. The first significant problem that this raises is that of time. This causes difficulties in a number of respects. Firstly, there will be a limited amount of time in which to make decisions concerning the optimum organisation structure for the project. People with a traditional perspective on the planning process tend to find this issue rather stressful, largely due to the fact that they lack confidence. This may seem a harsh statement, when in fact it is not. It is no reflection on their knowledge of the planning process, which may well put them in the expert category. The issue of confidence arises when individuals who are used to operating within the confines and constraints of a closed-system approach are suddenly placed in an open-system environment with fewer constraints; they are no longer sure where the goal posts are! Perhaps planners and managers are, by nature (or genes), more concerned with success than others and consequently are more needful of knowing how they are to be judged as being successful. One possible result of this is that they may consistently try to impose a closed-system approach onto an open-system project. Such actions are almost certainly going to result in conflict. These sorts of problems are covered in more detail in Chapter 4 where transformational organisation structures are introduced.

The second significant problem with open-system projects is the issue of 'mirage' projects. Due to the lack of constraints in such projects there is always the possibility that individuals will seek to extend the scope of the project. This may be in the context of adding new activities, further objectives or extending the intended lifetime of the project. In these ways, a false or mirage project begins to emerge in competition with the real project. Mirage projects tend to require changes to the original organisation structure planned for the real project. However, this can seldom be achieved in an explicit manner as those involved may lack the authority to introduce such changes officially. The tendency is for individuals to 'divert' resources on the basis of their unofficial authority within informal structures that may be largely invisible to those outside them. Wenger (1998) identified the existence of such structures in his work on communities of practice and found that to all intents and purposes they were invisible to those who were not members. This is not to say that such communities and structures are invariably detrimental to the organisations within which they are embedded. There is evidence that they can be, in many instances, the means by which tasks are completed when the official structure is incapable of achieving this. Emmitt (2001) found that gatekeeping (which it is suggested could be regarded as forming a community of sorts) can produce both beneficial and detrimental results for an organisation. The issue of pathways, with regard to both selection and maintenance, can therefore be a significant factor for a project and is worthy of consideration when developing an appropriate organisation structure. The key terms in this chapter, along with other terms from elsewhere in the book are defined in the glossary included as Appendix 2.

2.10 Conclusions

This chapter has covered quite a wide range of subjects, some of which will doubtless have caused at least the occasional raised eyebrow. In some cases, the reaction may have been more demonstrably negative. The discussion of the second law of thermodynamics in the context of projects and their organisation structures could quite conceivably be dismissed as being both subjective and without credibility. Such a reaction, however, points to perhaps the key conclusion of this chapter; the rigid mindset that has developed over centuries of organising for projects on the basis of transactional relationships and the structures that are required to support them. This is perhaps understandable, in the historic sense at least, in that success seemed to flow from control. By being able to control the production processes so as to reduce uncertainty, the probability of success was increased. Once this link was made, it perhaps became inevitable that more opportunities to impose control were sought. After all, at least as far back as Vitruvius project managers were attempting to reduce the probability of failure through relatively simple control measures such as specifying those sources of materials that were in some way 'better' than others. Quarry 'A' produces more durable stone than quarry 'B', for example. By using stone from quarry 'A', you will increase the probability that your building will last longer, and therefore be more successful. More successful buildings provided other opportunities, and technology advanced, resulting in more opportunities to exert control. As it became possible to analyse metals, for example, stronger and lighter components could be made.

Throughout this process of development the underlying message has been that not only can we control methods and materials, we can also control environments, both internal and external to a project or organisation. In this manner, it seems inevitable that a deterministic approach should have become the norm, and there is no doubting that for several hundred years it has produced significant advances. It is also arguable that those advances have not been without their costs. The opportunities missed through the stifling of creativity will remain forever unidentified, thereby preventing the true costs ever being established. While there are important ethical and moral issues within such a debate, that debate largely lies outside the scope of such a book as this. However, the advances previously delivered by the transactional approach have brought society to the stage of creating technologies that are increasingly capable of operating within open-system environments, such as the latest generation of robots. There is also the issue of individuals who are increasingly able to exert authority based on their knowledge rather than their position within a hierarchy. These technologies and individuals are an illustration that the traditional, transactional approaches to organising project structures are becoming less and less relevant to the sort of society that has developed over the last 40 years or so. Such an ongoing change suggests that at some point the transactional approach will no longer be able to meet society's requirements regarding the effectiveness of projects (which will also change). The question then becomes one of can the transactional approach be improved, or should it be replaced completely by a new approach? Pirsig (1974) raised similar concerns when he referred to the tendency of individuals to regard government and establishment institutions as 'the system' as being correct, due to such organisations being founded on structural conceptual relationships. These relationships, however, tend to support the organisations long after they have lost all other meaning and purpose, with the result that people carry out totally meaningless tasks and never question why. Pirsig argued that this scenario was without a villain forcing people to live meaningless lives; it is simply the structure that demands it. The true culprits are perhaps those who are unwilling to attempt the huge task of changing that structure, even when they realise it is without meaning.

PART II

STRUCTURE PRESENT

3 ESTABLISHING A PROJECT'S RELEVANT ENVIRONMENTAL FORCES: RECOGNISE THE RELEVANT – IGNORE EVERYTHING ELSE?

Certum est quia impossibile est – it is certain because it is impossible.

Introduction

Chapter 2 suggested the possibility that projects may not in fact be as amenable to the imposition of the control function as has been traditionally believed. If the assertion that some projects may actually be non-linear in nature proves to be correct, there is little point in trying to control the performance of such projects by seeking to enforce conformance with some predetermined model of how they *should* unfold. Within such projects the traditional emphasis on data-gathering and feedback may indeed represent a significant waste of resources (this suggestion will be returned to in Chapter 5). Similarly, if the suggestion that projects are essentially open systems and therefore able to respond to changes in their external environment is accepted, there is also a need to consider the project organisation's approach to data-gathering. In both scenarios the question is one of what data/information should be gathered and what should be ignored.

3.1 Gathering information

On the question of what information should be gathered for a particular project, two continua are suggested for initial consideration:

- *Mitigable – unmitigable*: the extent of a project team's ability to control or moderate adverse environmental effects. In order to quan-

71

tify information of this nature the organisation structure/system being proposed for a project must be examined.
- *Definable – undefinable*: the extent of information available on the probable effect(s) of a particular environmental factor.

(Moore & Moore 1997)

The first continuum may appear to conflict with the suggestion that projects may in fact be non-linear and therefore not susceptible to control in the traditional manner, but this need not be the case. Only if it can be fully determined that a project is in fact completely non-linear in nature should the possibility of ignoring this continuum be considered. The basis of this argument is that if the project team identifies the possibility to prevent a particular adverse environmental factor having any input into the project system, it should plan on making use of that possibility. If nothing else, it removes one interdependency from the complex interactions possible within a project environment, which is arguably worth doing even if the project is fully non-linear as it potentially makes life simpler.

Likewise, there are benefits to being able to determine the probable effect of a particular environmental force on the project. The mere fact that this proves to be possible for a given project indicates that it may not be fully non-linear and that there is some scope for intervention in the unfolding of the project. Even apparently minor interventions can have a significant impact on project performance and outcomes by the time they have passed through those project subsystems that are essentially non-linear. Of course, there is always the concern that the lack of a linkage between inputs and outputs in certain areas of a project may result in unanticipated outputs. However, by concentrating the data-gathering exercise on those environmental factors that can be clearly identified as having the potential to significantly affect the project in an adverse manner, any interventions will at least reduce the level of adversity and can therefore be regarded as being beneficial.

In both cases, there are implications for the project organisation structure. If, for example, there is no possibility of carrying out any interventions for which the outcomes can be identified with an acceptable level of probability, there is little point in implementing a control-oriented project organisation structure. However, in those cases where such interventions are possible, the project organisation structure should be such as to allow those interventions to be made at the correct time. The need to make this decision places considerable emphasis on being able to identify adverse environmental forces for a given project.

3.1.1 Adverse environmental forces

The possible range of adverse environmental forces is extensive, from the mundane (inappropriate filing systems) to the spectacular (civil war). The filing system example may seem to be a laughably simple one – after all, almost any filing system can be made to work if people are sufficiently committed to it. However, there are two points to consider within such a statement:

- the manner in which human resource energy is released (see Chapter 2) in completing project change; and
- the application of gatekeeping.

The concern with energy release is simply one of optimising the use of energy – if energy is being squandered on making a filing system work, there will be less energy available to deal with an unexpected project change event when it comes along. The possible application of gatekeeping is potentially a more insidious problem in that the project manager may not be aware that the problem exists. Emmitt (2001) identifies one form of gatekeeping within architectural practices with regard to the handling of information concerning new, possibly innovative, products.

Perhaps the most common route along which new product information from the external environment is imported into construction design offices is through the trade literature mailed to them. Emmitt found that up to 80% of the literature received in this way was rejected (simply thrown in the waste bin) within a period of about five minutes every morning. The criteria used for this selection were largely subjective and the majority of literature was discarded on the basis of the front cover's impression on the reader or on a perceived lack of 'good' technical information. Within the process there was an underlying exercise aimed at risk management – unknown suppliers were perceived as representing more of a risk than those that were known to the office. Some offices went further in this than others, in that they followed up the first cull with a second in which a different partner in the design practice perused the remaining 20% and typically reduced that by half, leaving only 10% of the original quantity.

One practice was cited as having a particularly elaborate system involving presenting technical information on selected new products to a panel of partners who then accepted or rejected the product information. Once accepted, the information was placed in the practice's library and formed part of the official technology palette from which all designs had to be developed. The reason for this approach was still one of seeking to reduce the risk of product failure and resultant legal action by clients. This strategic risk management approach by the senior partners carries, in effect, a creativity cost flowing from the restricted awareness of junior partners with regard to new products. In such a scenario the question could be asked as to which is the more adverse environmental factor – the restricted palette of technology with its impact on design innovation, or the risk of legal action from failed technology within an innovative design?

Guidance on the issue of factor identification in general can be found in the formalised approach within the work of Hughes (1989), which identifies 11 environmental factors (referred to as environmental subsystems) with a possible influence on projects. These environmental subsystems can be regarded as a significant widening of the previously discussed PEST/STEP (social, technical, economic, political) model of the external project environment. The 11 subsystems are:

- cultural;
- economic;
- political;
- social;
- physical;
- aesthetic;
- financial;
- legal;
- institutional;
- technological; and
- policy.

Hughes considers all of these in terms of the two continua mentioned previously (mitigable/unmitigable and definable/undefinable), resulting in an apparently comprehensive model of the project's external environment. Having identified a number of possible factors, the issue then arises of how to quantify the probable effects of any adverse environmental forces that are found within them.

3.1.2 Probable effects of adverse forces

Quantifying the effects on a project of relevant environmental forces, irrespective of whether they are beneficial or adverse, can be so difficult as to be near impossible. Chapter 2, for example, introduced the concept of non-linearity in project activities, and that concept will be developed further at this point. In completely non-linear systems, both the type and level of outcome(s) may be unquantifiable; it may not even be possible to identify whether the outcomes themselves are either broadly adverse or beneficial. In highly complex, rapidly evolving environments where data-gathering cannot be completed at sufficient speed, the outcomes may have been 'delivered' even before the project team is aware of the activities spawning them. Trying to deal with an ongoing extreme natural disaster is one example of this type of environment.

Nonetheless, humans persist in trying to plan and control highly non-linear projects, generally by attempting to impose linearity on non-linear systems. These attempts can be categorised as being based on either some form of statistical analysis (PERT, etc.) or on the expertise of one or more individuals. The former may be regarded by many as being reassuringly quantitative and therefore objective (not strictly true), while the latter frequently requires project teams to take a leap of faith which may be eased by an individual's perceived sapiential authority or leadership qualities. Military organisations, for example, tend to rely almost entirely on their hierarchies for effective operations during peacetime; one individual's positional authority is frequently sufficient to ensure that other individuals and groups will carry out the required actions. During states of emergency it is arguable that the importance of positional authority diminishes and that military organisations move to relying more on the sapiential authority of their commanders. In such environments the outcomes tend to be condensed into two direct opposites: safe (beneficial) and unsafe (adverse). Put simply, if the troops perceive their commander as having sufficient expertise (which we can regard as being broadly aligned with the concept of sapiential authority) to make a safe decision, they will follow his or her commands with little or no hesitation. There is seldom time on the battlefield for a focus-group meeting to determine commonly acceptable statistical values for probable outcomes of various possible actions!

It is worth noting that the emphasis in this section is really on identifying what is probable as opposed to what is possible. You may believe, for example, that there is a possibility that Elvis will be returned to Earth, on his next birthday, by the alien race that has been worshipping him for all these years. But even if such an event was possible, how probable would it be? Perhaps unfortunately, the human race does not always make decisions based purely on the probabilities of particular outcomes being delivered. It has been claimed that the average British resident has a greater probability of being hit by a meteorite than of winning the national lottery jackpot. Nonetheless, how many people do you know who have developed a personal meteorite protection strategy, as opposed to the number who have developed a strategy for spending the millions they are going to win on the lottery this week? On the basis that it is usually feasible to differentiate between the possible and the probable (albeit on the basis of personal knowledge and data gathered), it should then also be feasible to place the probable outcomes at a point on some of the more broad-brush continua, such as the favourable-hostile continuum.

3.2   *Application: a hypothetical project scenario*

The hypothetical project scenario is provided as a means of exhibiting the detail that can be achieved in determining the extent and strength of relevant environmental forces within a project. The scenario is developed from one previously introduced by Moore and Moore (1997). Key characteristics of the scenario are:

- client is an established and reputable company;
- client requires a state-of-the-art prototype wind turbine;
- development site is in a historic UK city noted for adverse climatic conditions;
- a rise in interest rates is seen as being probable;
- UK wages and prices are rising, but this trend is not being repeated in the rest of Europe;
- society has become increasingly litigious as a response to dissatisfaction with insensitive projects, and project management companies increasingly find themselves as defendants;
- management contracting has been selected as the procurement method;
- the selected management contractor is a young company with limited negentropy (negative entropy); and
- wind turbine designs are not complete as work starts on site preparation.

The above scenario must then be considered in terms of the environment, both external and internal, to the project. One technique for this consideration was proposed by Hughes (1989) in his examination of environments and was introduced earlier in this section. The technique was to quantify information about a project's environment under the two types of continua introduced earlier: mitigable – unmitigable and definable – undefinable.

This fairly basic process of considering a project's environment can be added to by incorporating Mintzberg's (1979) work on the favourable-hostile continuum, and also Duncan's (1971) work on both the static-dynamic and simple-complex continua. The resultant model of the project may seem to be particularly rigorous, but it should be remembered that it is still based on a largely traditional (that is, transactional) perspective of projects and project management. An important point to consider with regard to this traditional perspective is that it does not always appreciate one fact: that as information regarding the project is gathered within each of the above continua, the degree of openness required of the project's organisation structure will need to increase. This will be due to the project increasingly relying on, and interacting with, its external environment. It is possible to simply impose a highly mechanistic project system. Such a system will require little interaction with its external environment and will continue to function so long as throughput is permitted and both maintenance and regulatory activities are carried out. However, the probability of such an approach succeeding in optimising the operation of the project is suggested to be slim.

In order to carry out the consideration of a project's environment in a robust manner, a structured approach is required. This will invariably result, as far as the contingency approach is concerned (with its emphasis on trying to achieve a steady state), in the formalised gathering of information and the application of rules and procedures as the organisation seeks to identify and control all the points of interdependency between the project's internal and external environments. Unfortunately, the more ardent contingency theorists may not find even this model sufficiently comprehensive. For such individuals, it is possible to extend the analysis further by accepting that environmental subsystems operate not only at different levels (such as favourable – unfavourable) but also with different intensities. Differing levels of intensity were identified by Osbourne and Hunt (1995) in relation to a project's immediate (micro) and wider (macro) environments. Moore and Moore (1997) proposed a number of revisions to the model developed by Hughes to calculate the environmental factor for a given project. The revisions were as follows:

- Whilst Hughes' original environmental subsystems are retained, the criteria upon which they are assessed are revised to reflect the various continua that can be used to determine an organisation structure's degree of openness.

- A weighting provision has been included in order that the relative potency of each of the environmental subsystems within varying project circumstances can be reflected.
- Assessment values are attributed to the environmental classification continua rather than to the environmental subsystems.

The second of the above revisions resulted from the need to improve the realism of the model in terms of environmental subsystems operating not only at different levels but also with differing intensities (degree of influence) within varying project circumstances. Consideration of different levels of operation results from the work of Walker and Kalinowski (1994), which considered the immediate (micro) and wider (macro) environments. Further development of the model resulted in the suggestion of a third level, referred to as the meso level. This can be added in cases such as the hypothetical project, where procurement methods result in the appointment of contractors who are argued not to be entirely controllable by the project but are contained within it (Moore & Moore 1997). The nature of the suggested relationship between each of the levels and the single continuum of mitigable – unmitigable is illustrated in Fig. 3.1.

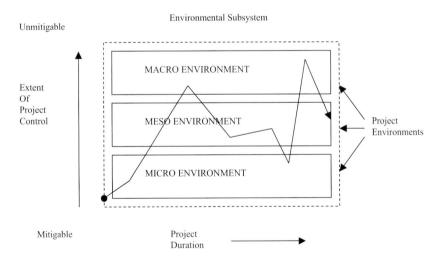

Fig. 3.1   Project environments and levels of control.

3.2.1 Project placement on environmental continua

The model takes each of Hughes' 11 environmental subsystems (only six are provided as examples in Table 3.1) and considers their effect(s) on the project from the viewpoint of the five closed–open continuums, with the effect being scored for each continuum as illustrated below:

- 0, subsystem tends towards the closed end of the continuum.
- 1, subsystem tends towards the centre of the continuum.
- 2, subsystem tends towards the open end of the continuum.

Following on from this, each subsystem is then weighted in terms of its influence over the project:

- 0.5, subsystem exerts relatively little influence over the project.
- 1, subsystem exerts moderate influence over the project.
- 2, subsystem exerts considerable influence over the project.

Within this weighting system it is possible to reflect situations such as that when a given subsystem exists within the macro environment of a project but still exerts considerable influence on the project, as is the case with Health and Safety Regulations. Likewise, a subsystem may be complex but also highly defined, in which case the profile achieved across the five continua would be different from that achieved by a complex but highly undefined project. By including a third level of environment (meso), a further possibility arises. This is the examination of the effects of organisations that nominally lie in the project's external (macro) environment but would normally be artificially placed in the internal (micro) environment through contractual arrangements, etc. This adds to the realism, in true rational-mind style, of the simulation being constructed.

By examining each of the subsystems in turn, a score is computed for each of the continua, with the five totals and their mean being plotted along the open–closed scale. This then provides the basis for determining the extent of openness required in the project's organisation structure. Sample assessments for the hypothetical project outlined previously are given as Table 3.1, and the total scores for each continuum, along with the overall mean score for the project, are presented as Table 3.2.

Table 3.1 Example scores on six (out of 11) sample environmental factors for a hypothetical project's closed-open assessment.

Environmental factor	Description	Weight		Simple–complex	Static–dynamic	Favourable–hostile	Defined–undefined	Mitigable–unmitigable
Cultural	Social attitudes to an organisational system's behaviour	0.5	*Raw Score*	2	0	1	1	0
			Weighted	1	0	0.5	0.5	0
Economic	Concerns general economic activity	2	*Raw Score*	2	0	1	1	0
			Weighted	4	0	2	2	0
Political	Government policy and its effects	1	*Raw Score*	1	0	1	2	2
			Weighted	1	0	1	2	2
Social	Concerns stakeholder views on the project	2	*Raw Score*	2	0	1	2	1
			Weighted	4	0	2	4	2
Physical	Topography, obstructions, hazards, weather, etc.	2	*Raw Score*	1	2	2	1	1
			Weighted	2	4	4	2	2
Aesthetic	Views on 'good taste' regarding the product	1	*Raw Score*	1	1	0	0	0
			Weighted	1	1	0	0	0

Table 3.2 Environmental continua; total scores (all 11 environmental factors) per continuum, and mean score.

Environmental continuum	Score (maximum total = 44)
Simple–complex	24
Static–dynamic	14
Favourable–hostile	21
Defined–undefined	17.5
Mitigable–unmitigable	19.5
Mean score	19.2

3.2.2 Diversity within the project

The model as discussed has shown that it is sufficiently detailed to allow a realistic consideration of the key subsystems within a project prior to the designing of an appropriate organisation structure. Analysis of the results indicates that the system representing the hypothetical project is not at either the closed or open end of the continuum, and that it is in the central area where it can be classed as a homeostatic system (one capable of a certain level of internal adjustment in order to subjugate the effects of the external environment) whilst tending towards being open. Due to the hypothetical project's representative system being assessed as more open than closed, the project therefore requires an organisational structure that is designed with an emphasis on flexibility rather than efficiency.

The above results should be considered in terms of the work by Shirazi *et al.* (1996), which suggests that when environmental conditions move towards becoming unfavourable for a given project, the parent organisation typically begins creating protective buffers around the project. This may be the case particularly when a project operates in a politically or environmentally sensitive environment. It can be argued that in such circumstances it becomes increasingly valid to consider the existence of a meso-level environment, in addition to the usual macro and micro environments, within projects. However, it can also be argued that adding levels of environment is simply an attempt by transactionalists to overcome the boundary problem identified by transformationalists (those who support post-contingency theory) and that such an attempt will ultimately prove to be counterproductive. One response to such an argument can be found in an examination of two further perspectives on organisation: behaviour and theory.

3.2.3 Problems with the continua model

The continua model discussed in the previous sections can be argued to be a more accurate model than its predecessors in that it allows for the inclusion of further detail concerning the project and its environments. However, this is not to say that it is the optimum model – optimised perhaps, but not the optimum. There are a number of ways in which the model can be criticised, and it may well be that addressing these criticisms would result in further optimisation of it. It is also possible that the energy required to further optimise the model may be better expended on the development of a different model. In order to try to resolve this potential dichotomy of development resources, we will now review the more significant problems regarding the operation of the continua model.

The first significant factor concerns the variability of factor assessment. This is primarily an issue of the impact of time on the project. As the project unfolds, it is quite possible that each of the factors used may migrate between the different levels of environment. In the early stages of the project's life-cycle, a particular factor may have a strong influence at the meso environment level, while at a later stage it may move to the micro or macro environment levels. While the ability of the model to recognise and accommodate this can be argued to add to its accuracy, it also raises a problem concerning the ability to forecast this type of movement. The basic argument goes along the lines of questioning the validity of the results from any model unless they are based on sufficiently accurate predictions of factor movements. Without the required level of accuracy being achieved, the model is arguably of use only in the here and now. Unfortunately, if the reasoning is taken further, the situation is actually worse than only being able to use the model to structure a project so that it can respond to the here and now. Arguably, the most accurate data concerning a project relates to the work completed rather than work-in-progress. It is therefore possible that models such as the continua one will, at best, give worthwhile results only if the data input is constantly updated or, at worst, can meaningfully respond only to what has already happened. This raises the intriguing possibility that perhaps we should not be seeking to impose structures on projects but that we should let them organise themselves.

The second problem involves the converging of assessments. This can be a problem when individuals are asked to make qualitative assessments of anything, particularly if they are not themselves of expert level. There is a tendency to converge or centralise assessments, particularly if the possible range is broken down into an uneven number of values (such as 0–4, 1–5, etc.). In this manner, all of the environmental continua will tend to converge towards the middle of the range. The inevitable result is that the project is deemed to be either partly open or partly closed, but in either case is not far from the midpoint on the continuum.

Again, the validity of the result from the model is open to criticism. Probably the most effective way of dealing with this type of problem is to give the task only to those people who are recognised as being sufficiently expert. Unfortunately, such people tend to already have had their expertise recognised and have had too many tasks imposed on them! An alternative is to provide all assessors with a clear-cut list of characteristics for each point on the assessment scale so that they can compare what they are presented with by the project to those characteristics. This immediately raises the issue of time required to operate in this manner. Perhaps this situation presents the possibility of introducing yet another functional specialism to the project team.

These problems are typical of the difficulties faced in gathering data for any tool intended to advise on suitable actions to implement, and the response to them is generally one of deciding to gather more (in terms of both quantity and accuracy) data. However, such a response can be argued to fall into the entropy trap as energy is wasted on non-productive tasks. Perhaps it would be more efficient to implement a lighter touch?

Memory test 3

Try the following questions:

(1) What were the two continua suggested by Hughes for environment assessment?
(2) Give 5 of the 11 environmental factors, or subsystems, suggested by Hughes.
(3) What are the two levels of environment identified in traditional systems models?
(4) What is the name given to a system in the central area of the closed–open continuum?
(5) What is the suggested response by parent organisations when conditions become unfavourable for their projects?

3.2.4 Required openness of organisation structure

The established belief on the study of an organisation and how it behaves is that such study can be split into two: organisation theory and organisation behaviour. Organisation theory actually concerns itself with the study of the manner in which a complete organisation and its major subsystems behave, while organisation behaviour concerns itself with the study of the behaviour of individuals and small groups within an organisation. The situation may be clarified by referring to organisation theory as the macro view of organisations while organisation behaviour is referred to as the micro view of organisations. Table 3.3 outlines some of the key differences between these two views.

While the common feature between both views is the importance placed on the people within an organisation, in the micro view their behaviour in terms of the communication, motivation, decision-making and leadership activities between individuals and within small groups is the most significant factor. Consequently, anyone designing an organisation from a micro viewpoint of organisation behaviour will emphasise the processes that focus on individuals and groups. The macro view of organisation behaviour, on the other hand, considers the most significant factor to be that of context: how people are aggregated into departments, divisions and organisations, and the structure (differentiation and integration, relation to external environment) and process (power, conflict, organisation life-cycle, etc.) issues within such divisions. In other words, the emphasis is on

Table 3.3 Micro versus macro views of organisation (after Banner & Gagne 1995).

Micro view (organisation behaviour)	Aspect	Macro view (organisation theory)
Structures and processes within individuals and small groups, and links between them	Issue emphasis	Structures and processes within major subsystems, organisations, and their environments, and links between them
Individual self-improvement and job design, intervention into interpersonal processes, training of leaders of small groups; individual and group change	Primary applications	Design and management of the structures and processes linking major subsystems, organisations, and their environments; organisational and environmental change

the organisation itself, and its relationship with the external environment, rather than the people within it.

Each of these views has its strengths and weaknesses, but both taken together can be deemed to represent what may be referred to as the traditional view of organisation study in which the collective behaviour of individuals (as groups and organisations) is seen as being a function of external influences (organisation technology, environment) and internal influences (power relationships, strategic decisions, organisation size). However, the postcontingency perspective on organisation study suggests that these two 'traditional' views are insufficient for the study of modern organisations and that they are lacking in an important area: patterns of social behaviour across groups of organisation stakeholders. Those who argue for this emphasis on social behaviour patterns also admit that the approach is still skewed towards the macro approach, in that it does not explicitly consider individuals other than as stakeholders in the activities of groups. There still seems, therefore, to be room for other ways to deal with the study of organisation, particularly with regard to the issue of those stakeholders who are not organisation employees.

It has been suggested previously that the traditional contingency theory perspectives on system environments may benefit from the addition of a third environment, the meso environment. Similarly, it can be argued that the traditional views on organisation study may benefit from the addition of a third view – yes, the meso view of organisation behaviour. This would result in organisation study being directed at three levels: macro, meso and micro. The previous suggestion was that micro study of organisation dealt with individuals and small groups, whereas macro study dealt with groups and organisations (and their relation to the external environment). Within this traditional approach, there seems to be an overlap at the group level, and it is at this level that it is suggested meso study could usefully be directed. Such an approach would result in the three levels of micro (individuals), meso (groups) and macro (organisations and the external environment), which, if nothing else, offers the opportunity for a faintly amusing acronym: IGO (Individual, Group, Organisation). The arguments for such an approach are fairly complex, but even so, it could still be said that such an approach simply represents an adjustment to the traditional approach to planning based on contingency rather than a significant step-change in planning project organisation structures, a point which will be developed further in Chapter 6.

3.2.5 Life-cycle models

The book has previously introduced the more common life-cycle models for projects (S curve, etc.) and have mentioned a team life-cycle. These two types of life-cycle can be brought together at this point as they are relevant to the issue of meso-level study, particularly if the meso focus of groups is taken to also include teams.

Along with the project life-cycle models previously covered, there is a model that considers the time distribution of project effort, as illustrated in Fig. 3.2. The relevant section of the model is the third one, dealing with effort related to planning, etc., as it is in this section that the greatest effort (or release of energy) is being, or should be, made. The possibility that the required effort will not be made is one that could be argued as falling clearly into the micro study level, in that it appears to be a problem of motivation. Even so, it could equally be argued that the problem is one for macro-level study in that, at this point in the project life-cycle, the project is primarily concerned with the issue of interdependencies with factors in the external, rather than the internal, environment.

While both arguments have a degree of validity to them, that is no guarantee that either is fully correct, and it is important that the issue of project effort is dealt with effectively, as it is possible that ineffective, or inappropriate, treatment will result in two unacceptable scenarios. The first is that individuals within the project do not commit to the project objectives and that groups containing those individu-

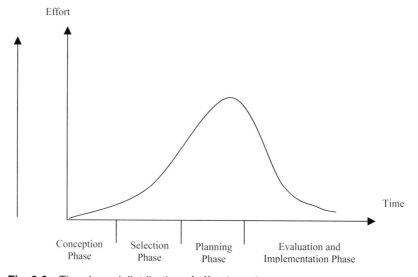

Fig. 3.2 Time-based distribution of effort in projects.

als make insufficient and/or inappropriately directed effort. Project performance then suffers. A second scenario is that individuals come together as teams who are over-committed to the project objectives, or at least the objectives that they see as being important. This can result in teams focusing on achieving what is in effect excessive perform- ance – a situation that some may have difficulty in seeing as being problematic. In the case of such individuals, they should consider for a moment the issue of planned delivery of resources to the project. When a team gets ahead of schedule, it is possible for the project to hit delays that were not identified in the original project schedule simply because they have to wait for other resources to be delivered. The frustration this brings can then cause significant problems (conflict) between team members and other project stakeholders.

Excessive performance actually delays the work being carried out and pushes up costs. This can be illustrated by considering the plan- ning function. There is a range of possible subactivities that could be dealt with under the planning function, all of which take time and cost money, so the more effort that is put into planning, particularly once the production phase of the project commences, the longer the project duration becomes as the original schedule is ignored in the search for better performance than was originally required. Meso- level study would be an opportunity to examine ways in which indi- viduals' levels of commitment could be matched to the requirements of the project life-cycle through management of the team life-cycle, one aspect of which is the achievement of teamthink. The concept of teamthink is covered in more detail in section 3.3.2.

3.3 Project objectives and organisation structure

Case study 2: The Yangtze Three Gorges Project

Introduction

The Yangtze Three Gorges Project (YTGP) has achieved a level of interna- tional recognition as a significant, complex and long-duration infrastructure project centred on the construction of a large hydro-electric dam. It has also aroused claims of being an environmental disaster in the making. There can be no escaping the conclusion that projects such as this one engen- der strong positive and negative opinions amongst their large number of stakeholders. For those who are not familiar with the project, a few facts and figures may be of use:

- YTGP is the largest water conservancy project ever undertaken by the People's Republic of China and is claimed to be the largest in the world, with a total storage capacity of 39.3 billion m³ of water.
- The site of the dam is in the Sandouping district of Yichang City, Hubei Province.
- YTGP will have a total installed generating capacity of 18,200 MW. This will allow the replacement of approximately 40 million tons of coal used annually in power station generation.
- The total water catchment area is estimated at 1 million km².
- Total length of the dam will be 2.309 km.
- Project duration is planned to be 17 years, with completion due in 2009.

The primary intention of the project is to control the severe flooding that has historically affected this section of the Yangtze River. Flood control ability at Yichang, for example, will be increased from dealing with a once in 10 years flood level to a once in 100 years level.

3.3.1 Resource variability

Within a project such as YGTP, a number of problems will inevitably have to be addressed in the development of a project organisation structure. One significant problem may be referred to as resource variability, and it is significant because it can affect any resource, on any project. The nature of this problem is possibly most easily understood when considering two general types of resource: materials and labour.

Projects carried out in the so-called developed world tend not to appreciate the extremes to which samples of a given material may vary over the life-cycle of a project. This is perhaps because such projects have developed mechanisms focused on ensuring an acceptable level of consistency within the materials used. A typical mechanism is the use of a performance specification (defined in Appendix 2). The emphasis on the deliverable requires the planner to work backwards and determine what resources, in terms of both quantity and quality, will be required to achieve the required performance for a specific deliverable. This will usually involve the consideration of a number of alternatives, particularly if the specification has not unduly constrained the process by over-emphasising technical attributes beyond what is strictly required. However, it is not unknown for technologies to be simply imposed because of the conservatism, or lack of creativity, of decision makers. Moore and Ahmed (1997), for example, found that whilst developing countries could identify indigenous alternatives for some construction materials, imported materials (and attached technologies) were generally preferred, largely because of a perception that they were in some way superior, irrespective of any evidence to the contrary. Likewise, it may well be possible to build a significantly sized dam without the use of steel-reinforced concrete (an energy-intensive material to produce), but the Chinese authorities seem to have elected to use the standard material for this type of project. Material resource variability may well have been a factor in their deliberations, but it is more probable that this was more of a consideration when considering the labour resource than the material resource.

A survey by *Building* (Madine & Black 2001) found that there were different levels of demand amongst the construction professions in China. There seems to be a plentiful local supply of engineers, so wages tend to be low, but there is a shortage of consultants (due to the relative immaturity of China's still-developing construction industry) and these are being recruited from overseas with the offer of more lucrative employment packages. How a Chinese engineer would feel about working alongside a more highly paid non-Chinese consultant is perhaps just one example issue with regard to variability in the labour resource and leads nicely to the need to consider teams and tribes within the labour resource.

3.3.2 Teams and tribes

This section could also be titled 'teamthink versus groupthink', in that it is possible to regard tribes as being largely equivalent to groups. This is not to say that tribes are not capable of operating in the manner of teams, but is simply an acceptance that such a form of operating is not always their primary objective. Conformance, for example, is arguably more important for long-term survival in the tribal situation than it is for short-term problem solving in the team situation. As with the project life-cycle, teams also have a life-cycle and their effectiveness varies over it, as illustrated in Fig. 3.3. The peak of productivity for the team arrives towards the end of the resolution phase and continues through the synergy phase. It is during this period that evidence of teamthink emerges:

- the expression of divergent views;

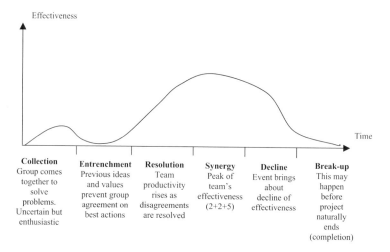

Fig. 3.3   Effectiveness profile of team life-cycle.

- expression of concern; and
- an awareness of limitations, etc.

Prior to this, the group (not yet a team) is characterised by evidence of groupthink:

- individuals within the group display considerable effort in trying to agree with each other; and
- any attempts to adequately discuss alternative solutions, etc. are suppressed.

Groupthink is typically seen as resulting in defective decision making and is therefore to be avoided within the project environment.

It would appear on the basis of the above that a good project manager should be seeking to ensure that the peak of the project team's effectiveness profile coincides with the peak of the project distribution of effort curve. Such an approach is made more difficult if membership of the project team does not remain constant, a particular problem with regard to the formation of integrated project teams (IPTs). Unfortunately, the problems do not end there. While there is the need for a group to develop into an effective team (in terms of reaching synergy), there is also the need to recognise that at different stages in the project life-cycle teams will typically be required to deal with different types of problem. There is therefore an argument that teams should certainly reconstitute themselves in terms of membership (different skills/functions for each new problem) and possibly also in terms of adopting different structures at different stages of the project. In this manner they may best respond to the type of problems typical of each project stage or phase. The YTGP is very much a long-term project and as such would be expected to deal with all changes in its external environment. However, due to the project being carried out in a nominally socialist country, this does not seem to have been considered an issue. The management structure is very much a traditional (transactional) one, as shown in Fig. 3.4.

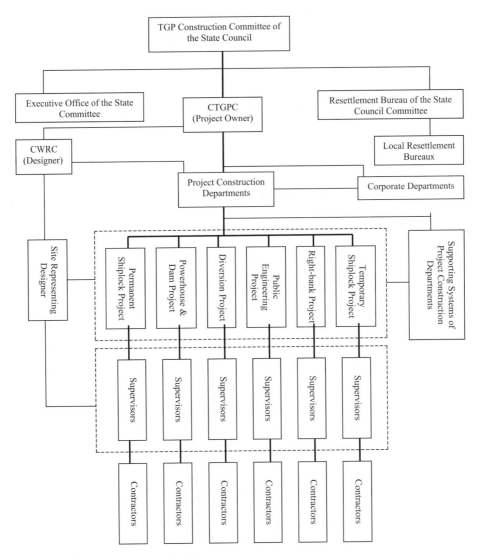

Fig. 3.4 YTGP management structure.

An example of a phase-responsive approach is the consideration that at the planning stage of a project, when project effort is starting to increase, the team will ideally need to be at its synergy phase. It will also need to adopt a category of structure referred to as 'creative'. During the execution stage of a project, the team will typically need to move to a 'tactical' category of structure, and also adopt a 'problem-solving' category of structure. Each of these categories requires different team characteristics to be in place. The creative category, for example, is characterised by a high degree of autonomy for the team members to explore the widest possible range of solutions. This can be contrasted with the tactical category where the team works to a well-defined plan, with high clarity of objectives. In order to achieve these differing characteristics, the team needs to have different types of people within it. Looking at the creative category again, this requires team members who are self-starting, independent thinkers. Managing the team development in the context of the requirements of project life-cycle phases can therefore be argued to be essentially a meso-level activity. The advent of the knowledge worker in modern society adds to this argument and also serves as a further example of the need to recognise relevant environmental forces for a given project. The relevance of the team to the issue of structure is developed further in Chapter 5.

Organisations which rely upon the contingency approach to manage their operations are increasingly finding that they have to deal with a new type of worker: the so-called knowledge worker. These individuals are typically defined as being people who know more than those who are above them in the organisation's hierarchy. They are therefore considered to represent a potential undermining of the power base of any individual whose authority is legitimised by their position within the organisation hierarchy. This again raises the concept of sapiential authority as differing from formal, or positional, authority – it is authority based on knowledge or expertise rather than on position in a hierarchy. Such a situation is increasingly regarded as being one of the symptoms exhibited by the demise of the industrial paradigm which has dominated the study of organisations since the advent of the Industrial Revolution. Failure by an organisation to recognise sapiential authority as a relevant environmental factor could impact negatively on the development of teams. This, in turn, would affect overall project performance. As far as YTGP is concerned, this could become a problem if large numbers of overseas consultants are ever expected to work within its apparently transactional structure.

The subject of groups and teams within project management is typically approached from the perspective of the bringing together of individuals having differing specialisms and expertise required by the project. These individuals are then developed into a team during the duration of the project. However, this is not the only reason why organisations use groups (and hopefully develop them into teams). Handy (1999) suggests the following additional reasons:

- management and control of work through organisation and control by appropriate individuals;
- problem solving and decision making through bringing together skills, talents and responsibilities having a capacity to produce a solution;
- information processing through passing on information to those who need it;
- collection of information, ideas and suggestions;
- testing, validating and ratifying decisions taken within or outwith the group;
- co-ordination of tasks and problems between functions/divisions;
- allowing and encouraging individuals to become involved and thereby increasing commitment to the organisation;
- negotiation and conflict resolution between divisions/functions/levels; and
- inquiry into the past.

Along with inquiry into the past, organisations are starting to be aware that developed societies are moving into the future, a future that it is suggested will increasingly refer to a post-industrial paradigm. How this will take place in a socialist environment such as China is not yet clear. As part of the move to a post-industrial paradigm, new ways of looking at organisations are required to deal with emerging factors such as knowledge workers. Of course, the acceptance of the validity of a meso level of organisation would have a repercussion on the traditional model of macro and micro study of organisation, in that meso study would have to be added. On the basis of micro study being labelled organisation behaviour, and macro study being labelled organisation theory, perhaps meso study would be appropriately labelled organisation intermediacy? This may be particularly useful when considering the possibility of conflict within an organisation, one possible cause of which is the determination of standards for the product(s) of a project and the activities within it.

3.3.3 Checking the product

At its most essential level, the process of checking the product can be argued to be about success and failure. If the product has achieved the required standard, the project team has been successful and everyone goes home happy. However, if the product does not achieve the required standard, the team is deemed to have failed and the witch-hunt for those responsible begins. For those who are given the task of working to achieve the standard set, the objectives should (if the design team has done its job well) be clear and achievable, so stress levels should be quite low as the team organises itself to achieve the stated standard. This is particularly so if the task-team is experienced and knowledgeable concerning the process(es) involved. On the basis of their knowledge, the task-team members should be capable of organising themselves with regard to who does what within the identified processes. There will naturally be some functional specialism within the task-team, although the extent of diversity regarding the types of specialisms may vary considerably between task-teams.

In the so-called scientific management model, with its emphasis on techniques such as work study and job evaluation, the tendency is to develop task-teams with minimal diversity as the task is broken down, excess functions are designed out and processes are redesigned in the search for greater working efficiency. This traditional, Industrial Revolution-era approach to organisation tends to produce highly specialised functions, along with the need to train the human resource so as to achieve a predetermined level of competency in these functions. Figure 3.4 provides an example of this approach within the YTGP project structure. YTGP is managed on the basis of clearly differentiated specialisms within a traditional hierarchical structure.

However, it is important to be aware that the training process in a transactional environment will give individuals a level of competency in, at best, a narrow range of functions or roles. This results in task-teams whose members have clearly defined primary roles but who may have few, if any, abilities with regard to secondary roles, and such task-teams are therefore compromised with regard to their ability to handle change. This is particularly so if the change happens rapidly.

Rapid change can be regarded as being problematic for industrial-era organisation structures. Their rigid, hierarchical forms do not reform themselves rapidly in response to environment changes and so are not suited to environments that may be classed as unstable or turbulent. This point will be illustrated in the next section, where traditional structures for project organisation are considered. However, the key point here is to consider the need for such structures to deal with the potential problem of change in their environment and their preferred strategy for this: control.

Traditional organisation structures seek to control their environment and thereby resist change. One device on which they rely heavily may be referred to as the quality control mentality. As with any other device, quality control has its relevance to production processes, but that relevance has to be maintained through an awareness of where it is, or is not, sensible to apply it. In the example of the traditional task-team structure, there may be plenty of opportunities to apply quality control, but not all of these will necessarily be relevant. The structure of the task-team will offer an opportunity to apply quality control at every point where product moves between functional specialisms within the overall process. At each of these points it is possible to identify a boundary (as in the systems model) where product can be checked as it exits one functional specialism and enters another. The question to ask is one of why there should be any need to check at either point, and the answer invariably comes down to a lack of trust.

There seems to be a certain level of irony in such situations – individuals who have been specifically trained to carry out limited specialisms to a specified level of competency, not then being trusted to carry out that task with the required competency. Obviously this will not apply in all situations. Pharmaceutical processes, for example, can be argued to place reliance on quality control activities for valid reasons of product safety. Irrespective of how knowledgeable those involved may be, mistakes or unplanned events can happen. Such an event may result in contamination of one product by another. This, in turn, could result in adverse, possibly fatal, reactions amongst consumers. Such a situation places great responsibility upon those involved in the production processes concerned and we all, as potential consumers, probably feel a considerable degree of relief that organisations respond to the responsibility by imposing reassuring quality control systems that are perceived to be based on specifications to achieve safe performance. In other situations it may not be the case that such consistent levels of safe performance are required and it is then worth questioning why a particular organisation structure that emphasises the quality control mentality is to be used.

In the case of YTGP, the issue of trust is an important one. The completed dam will be required to achieve a high level of performance and safety, and the risk of this being compromised by those involved in its construction will doubtless be seen by the project's client as something to be reduced where possible. There have been a number of building collapses in China recently and a recurring theme is that of corrupt contractors using inferior materials and construction methods, thereby compromising safety. A dam the size of this one needs to be provably safe.

Organisation intermediacy was previously suggested as tying in with the concept of a meso level of environment. If intermediacy is regarded as being the point at which organisation theory transmutes into organisation behaviour, the structure at this level has particular relevance for the functioning of a project. In the event that the structure impedes this transmutation, the functioning of the organisation (or project) will be impaired. Conflict is suggested as one example of this possible impairment, in that if the release of energy by the human resource is directed at dealing with conflict (in its harmful, rather than beneficial, form), it is being diverted from the processes where it was intended that it be used. Production therefore suffers and, once again, the witch-hunt begins to find those responsible for the failure. Project managers and others involved in the process of determining structure for a project therefore need to question how their decisions concerning the overall structure may impact on the effective operation of a given project. This may be particularly the case where it is standard practice within a parent organisation to simply impose a traditional form of structure (such as in the YTGP) on every project, irrespective of the needs of the project organisation.

3.4 Traditional organisation structures

There are two important issues to be addressed when initially dealing with organisation structures. Firstly, there are issues at the project level, with regard to how to determine the most appropriate project structure in order to achieve suitable group dynamics for making a cohesive whole (that should ultimately become a team) of various individuals/stakeholders. Secondly, there are further issues to be considered at the strategic level with regard to how the selected project organisation structure fits with the parent organisation structure. In multi-project environments, it is possible that each project environment may have to mesh with more than one parent organisation structure, thereby increasing the complexity of the structure selection function.

Such issues lead to the consideration of the role of projects within the three basic organisation structures: functional organisation, pure project organisation and matrix organisation, particularly the co-ordination model. However, there are several other 'standard' organisation structures that will also be introduced in this chapter. A further consideration is that an organisation's life-cycle suggests that as it matures it tends to increase in size, and as it does so the degree of individual specialism by stakeholders also increases. The traditional approach when setting up a new business, or parent organisation/ entity, has been to adopt the functional specialism or chimney structure, as shown in Fig. 3.5. This, however, has been shown to have disadvantages in that it restricts individuals' creativity and therefore the ability of the resulting organisation to respond to changing market and operational needs.

When an organisation seeks to involve itself in projects, there usually emerges the need for cross-function activity between the functional specialisms of organisation stakeholders. Work groups and teams therefore result from the situation where one specialism is seldom able to achieve the full range of a customer's or client's requirements. As the majority of projects are defined in terms of more than one function, the project manager is increasingly having to define organisational structures at the project level. It is therefore important that project managers are aware of the main possible structures. Organisational structures are also important because:

- they define responsibility and authority;
- they outline reporting arrangements;
- they determine management overheads (costs);
- they set the structure behind the organisational culture; and
- they explicitly determine the stakeholders in project activities.

(summarised from Maylor 1996)

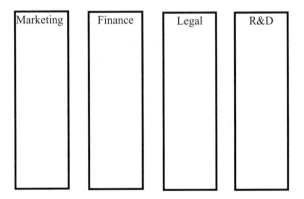

Fig. 3.5   Chimney organisation structure.

It is also important to realise that as organisations grow in size, the level of integration between functions becomes less and less. Basically, large organisations are not seen as being designed to integrate. This can be a significant problem for project managers given their key role in balancing differentiation in a project through seeking integration. However, there are various project organisation structures that attempt to overcome this problem without the need to revise the parent organisation structure. The success of these structures in dealing with this problem varies, thereby leading project managers to consider designing bespoke structures for each project. The following sections introduce the three most commonly used forms of transactional organisation structure.

3.4.1 Projects: functional organisation

In this approach, the intention is to give each project a home within one of the functional divisions of an organisation. The division is selected on the basis of which one has the most desire for the project to succeed and would therefore be expected to exhibit the greatest level of stakeholder involvement and motivation. With a project of a particularly differentiated nature, there may be no clear expectation that any one division would most benefit from it. In such situations the project manager is faced with a number of possible candidate homes from which the project can be administered, albeit with possibly little motivation for the project to succeed. Figure 3.6 illustrates an example of the functional organisation structure.

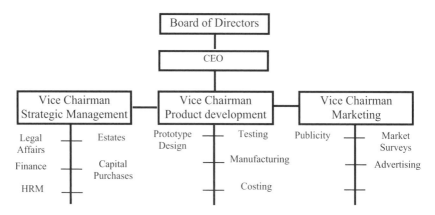

Fig. 3.6   Functional organisation structure.

If the correct division or function is chosen, the functional organisation structure has the following advantages and disadvantages for projects.

Advantages

- There is maximum flexibility in the use of staff.
- Individual experts can be utilised by many different projects.
- Specialists in the division can be grouped to share knowledge and experience.
- Functional division also serves as a base of technological continuity when individuals leave the project/firm.
- Functional division contains the path of advancement for those with expertise in the functional area.

Disadvantages

- The client is not the focus of activity and concern.
- Functional division is not usually problem-oriented in the manner that a project must be in order to be successful.
- Occasionally, no individual is given full responsibility for the project and disorder results, along with slow responses to client needs.
- There is a tendency to suboptimise the project.
- It tends to result in low motivation of assigned project staff.
- It does not facilitate a holistic approach to the project.

(summarised from Meredith & Mantel 1995)

In many respects it can be argued that the functional organisation is not the ideal for the management of projects. Nonetheless, it may be selected by organisations which spawn few projects, but the nature of those projects is that they involve either focusing on in-depth application of a technology, meeting a specific schedule or ensuring rapid response to change.

3.4.2 Pure organisation

This structure is generally seen as being at the opposite end of the spectrum to the functional organisation structure due its use of self-contained units for each project. Figure 3.7 illustrates a typical pure project organisation structure.

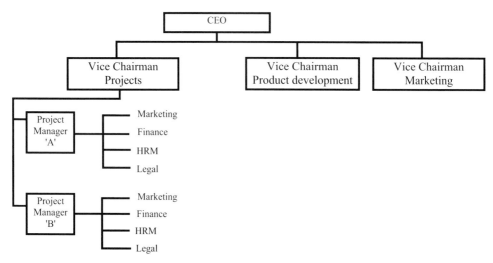

Fig. 3.7 Pure project organisation structure.

Individual projects will have their own staff and very few ties to the parent organisation. Examples of the sort of ties that may be imposed can be found by considering the minimum and maximum levels of tie. The minimum tie imposed is that of final accountability – so long as the project objectives are achieved, the process is of no significant interest to the parent organisation. The maximum tie imposed relates to prescribed procedures for administration, along with financial, personnel and control activities.

The advantages and disadvantages of the pure project structure are given below and should be compared with those of the functional organisation structure given previously.

Advantages

- The project manager has full line authority over the project.
- All project workforce members are directly responsible to the project manager.
- Removal of the project from functional division shortens lines of communication.
- It allows maintenance of a near-permanent group of expert project managers.
- It encourages a high level of commitment from team members.
- Centralised authority allows rapid decision making.
- There is unity of command– one, and only one, boss!

- It is structurally simple and flexible, making it easy to understand and implement.
- It tends to support a holistic approach to the project.

Disadvantages

- Fully staffed individual projects leads to duplication of effort.
- It tends to result in stockpiling of resources by the project manager.
- Technical experts can fall behind in technology developments outside their project.
- There are possible inconsistencies in carrying out procedures and policies.
- Projects may take on a life of their own (projectitis).
- There may be uncertainty amongst team members regarding employment after the project ends.

(summarised from Meredith & Mantel 1995)

While pure project structures have a greater number of advantages with regard to the management of projects (as would be expected), their disadvantages mean that they will rarely be adopted by parent organisations whose primary activity does not revolve around the completion of projects. The pure structure is therefore most suitable when a firm engages in a large number of similar projects, and also for one-off, highly specific projects requiring careful control while not easily being linked to a single functional division.

For those parent organisations which tend to operate somewhere in the middle ground where it is not fully clear that they should adopt either the functional or the pure project structures, there is a further possibility: the matrix structure.

3.4.3 Matrix organisation

This structure is an attempt to combine key advantages of the pure and functional structures. Effectively, it is a compromise achieved by imposing pure project organisation onto the parent organisation's functions. Figure 3.8 illustrates a typical form of the matrix structure.

In the matrix structure the project manager controls what is to be done by the individuals and groups assigned to a project, and also when it will be done. The functional manager within the parent organisation structure will decide which individuals or groups to

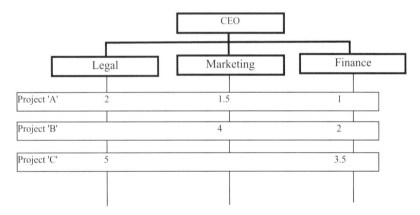

Fig. 3.8 Matrix organisation structure.

assign to a project and the type(s) of technology to be used. Project managers who are considering the use of a matrix structure need to take into account a number of factors. For example, success for each of the various forms of matrix structure (see following sections) depends upon:

• training on working in matrix environments – this should be given to both managers and team members;
• the quality of administrative, informational and career support systems; and
• an individual's nature – most important is their tolerance for role ambiguity brought on by conflicting priorities between the department and the project.

Disadvantages for matrix structures include:

• anarchy – the perception that as soon as something is working it is changed;
• groupitis – an individual will not make a decision without group approval;
• stifling – the time required to achieve group consensus stifles individuals' imaginative flair and the group becomes a barrier to rapid progress; and
• cost – excessive cost overheads can result.

However, matrix structures do have a number of advantages and have been found to work well for some organisations in certain circumstances. Typical advantages include:

- the project is the point of emphasis, where an individual takes responsibility for managing the project;
- the project has reasonable access to the entire reservoir of technology in all functional divisions;
- when several projects are in progress at the same time, the duplication found in the pure structure is significantly reduced; and
- response to client needs is as rapid as in pure structures. The matrix structure also responds rapidly to demands made within the parent organisation. A project nested within an operating firm must adapt to the needs of the parent firm or the project will not survive.

The level of success achieved by matrix structures depends upon the success criteria established for a given project. Team-based structures perform better on time and cost criteria, whereas purely matrix structures perform better on quality (technical specification) criteria. The matrix structure can therefore be concluded to be, in terms of transactional thinking, the only real choice for projects covering several functional areas that involve reasonably sophisticated technology but do not require consistent input from all the technical specialists.

In attempting to make the matrix structure effective within a range of projects and parent organisations, various forms of the structure have been developed and are available to the project manager. Three such forms are the co-ordination model, the overlay model and the secondment model.

Co-ordination model

The project manager acts as a co-ordinator for the project and chairs meetings of representatives of all departments involved in the project. Responsibility for success of the project is shared between all departments involved. This model is generally considered to be the most ineffective form of matrix structure due to the functional managers having greater power than the project managers. As a result, project meetings can be distorted by the functional managers.

Overlay model

This model attempts to balance the power of the project manager and the line manager by requiring the project manager to compensate the line manager for the temporary loss of resources. This is done through the project contributing towards the function's income. Unfortunately, this results in the creation of a second line of command in which one person will have both project and line responsibilities. This is generally seen as being a significant disadvantage of the model.

Secondment model

Within this model the function departments provide resources (individuals or groups) through full-time secondment to the project team for the duration of their required involvement in the project, after which they return to their line function within the parent organisation. This allows the costs of specialists to be incurred only as those specialists are required. However, this approach causes problems for individuals through the creation of task discontinuity as they move between their role within the parent organisation and that in the project.

Memory test 4

Here are a few more questions in order to get the memory working.

(1) Give three advantages and three disadvantages for the functional organisation structure.
(2) What is the minimum tie that a parent organisation can impose on a pure project structure?
(3) Give four advantages and four disadvantages of the pure project organisation structure.
(4) In what way can a matrix structure be regarded as a compromise?
(5) Give three advantages and three disadvantages for the matrix organisation structure.

3.5 Variations on the basic themes

In addition to the functional, pure project and various forms of matrix organisation structures there are a number of other variations on the basic themes of structure. Each of these variations is claimed to be suited to particular circumstances. The multinational structure, for example, offers several options to fine-tune its basic objective to maintain three-way organisation perspectives and capabilities across the factors of products, functions and geographic areas. Not the sort of structure that immediately springs to mind as being suitable for project management. However, if the project under consideration was one with the objective of opening up a new market in a particular geographic location away from the parent organisation's location, the multinational structure might well be suitable. The variations to be covered in this section are place, multinational and network structures. These variations are included primarily to give an overview of the diversity of organisation structures available, rather than to suggest that any of them are particularly suitable, or unsuitable, for use as project organisation structures.

3.5.1 Place structure

This approach involves identifying an organisation's primary units and then establishing them geographically (a particular region or territory) while retaining aspects of functional organisation structures. The result tends to be the placing of a range of tasks under the control of a single manager rather than taking the functional approach of managers dealing with single, specialist functions. Claimed advantages of this approach are that it allows organisations to deal effectively with regional cultural and legal differences when the parent organisation is operating across a number of countries/geographic markets. This is a particular benefit for organisations whose primary activity deals with items such as pharmaceutical and healthcare products, although the American Internal Revenue Service also uses place structures.

A further advantage of place structures is that divisions are presented with the possibility to become more readily aware of changes in the needs of their customers/clients and therefore respond more rapidly to such changes. However, a possible disadvantage is that control and co-ordination problems increase, and there is the possibility that regional units/divisions may begin to diverge significantly, with resultant problems for integration of the whole. Other disadvantages potentially include:

- divisional managers may wish to control their own resources and internal activities;
- employees begin to emphasise their own division's needs and goals rather than those of the parent organisation.

Again, this is not a structure that presents obvious advantages for project management.

3.5.2 Multinational structure

As mentioned previously, the multinational structure offers several options to fine-tune its basic objective to maintain three-way organisation perspectives and capabilities across the factors of products, functions and geographic areas. Unfortunately, in order to attempt to achieve a perfect balance between the three factors (which is generally seen to be impossible), a three-way matrix structure would be required. Given that the earlier discussion of matrix structures considered only a two-way matrix, this should indicate the potential for difficulties with a three-way version. In order to avoid such difficulties, multinational structures emphasise only two of the three factors. Typically these are place and product design structures, with functions being relegated to third place.

However, the degree of emphasis on place or product design structures individually can create tensions within the parent organisation. For example, a strong emphasis on place structure will give regional/divisional managers a high level of authority which will allow them to respond to regional variations. Emphasis on product design structures, on the other hand, gives product line managers a high level of authority, which they may use to seek out global efficiencies through greater integration, such as can be achieved with standard products.

An example of multinational structure emphasising place structure within the function of product design can be found in Ford cars. When faced with the challenge of reducing its product development times and costs, the company found that moving from a place structure based around quasi-autonomous regions to a product structure which acted globally reduced costs by around 30% and development time by at least 25%. This new approach allowed Ford to develop a range of 'standard' cars that could be marketed globally by making relatively small regional variations to their specification.

3.5.3 Network structure

The network approach to designing organisation structures aims to overcome the disadvantages faced by other structures with regard to dealing effectively with rapidly changing (turbulent) technology and environments. Network structures are claimed to be suitable for the managing of very diverse, complex and dynamic factors that may be external or internal to the parent organisation. This is achieved by the structure seeking to focus on the sharing, rather than allocating, of authority, responsibility and control amongst the multiple units and individuals who must co-operate and communicate frequently with each other in order to achieve goals held in common. Network structures are also claimed to be sufficiently versatile to allow movement between various options within the structures in order to accommodate change.

A typical network structure diagram is generally regarded as being somewhat reminiscent of a pepperoni pizza. The round shape of the pizza is intended to illustrate that nobody within the structure is more important than anyone else. The slices of pizza represent the areas where collaborative interaction between the pepperoni slices (larger teams, cross-functional in nature) takes place. The pieces of sweetcorn (although there is probably no problem in having olives if you don't like sweetcorn) represent either the smaller support teams dealing with issues such as human resources or one-off special project groups that will be disbanded upon completion of their objectives. Other names for the network structure are spider-web and cluster (not to be confused with the cluster structure suggested by Handy (1999): see Chapter 4), names suggestive of the proliferation of interdependent mechanisms and managerial processes they contain. Because of this proliferation, it is particularly difficult to represent the network structure using normal organisation charts and diagrams.

This section rounds off with the conclusion that there are many organisation structures to choose from and as a consequence the project manager can be faced with the difficult process of trying to select the most appropriate structure for any forthcoming project. However, before introducing how the selection of a structure may be approached, look back over the various structures covered and determine which one, if any, matches up with the structure (or structures if you are working in a multi-project environment) you think you are working within currently.

3.6 Selecting an organisational form

This is invariably a difficult area. The choice of structure for the interface between the project and the parent organisations is largely determined by the set of circumstances prevailing at the time. However, as with most areas of expertise, the selection process is partly intuitive. Perhaps as a consequence of this, there are no detailed procedures generally available for determining what form of structure is needed for any given project. There may well be some in-house guidance available within some organisations, but this rarely makes its way into the public domain. Such a situation encourages the project manager to consider the probable nature of a project, compare this with the advantages and disadvantages of each structure option, consider any cultural preferences of the parent organisation, and ultimately reach the best compromise possible under the circumstances! This is sometimes referred to as the process of optimisation: the result is not perfect, but it is the best that could be achieved at the time. This optimisation process can be formalised to some extent into a six-step procedure as laid out below.

3.6.1 Six-step optimisation procedure

Step 1. Determine the kind of work that must be accomplished. This requires an initial plan identifying, for example, the primary deliverables for the project and who is responsible for each deliverable. This information may be presented as a table such as:

Primary deliverables	Functional unit responsible for task
Deliverable 1	Unit 'A'
Deliverable 2	Unit 'B'
Deliverable 3	Unit 'C'

etc. This table can then be developed to produce a statement of objectives, identifying the major outcomes desired (each typically around 2–10% of the total project) for each of the primary deliverables identified.

Step 2. Determine the key tasks required to achieve each of the objectives stated.

Step 3. Arrange the key tasks into a work sequence, then decompose each task into a work package.

Step 4. Determine which project subsystems are required by each work package and which subsystems will work most closely together.

Step 5. List any known special factors concerning the project – duration, etc. Include any previous experience(s) of the parent firm/organisation with different project structures (also identify any cultural preferences).

Step 6. Select the most appropriate project structure!

(summarised from Meredith & Mantel 1995)

It is quite probable that the selected structure will be one of the three basic structures introduced previously: functional, pure project and matrix. Note that the selection procedure becomes somewhat more complex when the project manager is operating as an external consultant to an established firm or organisation. This is not to say that the problems are insurmountable, but it helps if you are the sort of person who enjoys a challenge. For those who would like the opportunity to get some practice in the area of selection, try the examples given in Meredith and Mantel (1995), pp. 167–170.

3.7 Integrated project teams (IPTs)

The concept of integrated project teams (sometimes referred to as integrated programme teams and integrated product teams) is one that is used in a number of areas to overcome the interfacing problems that can exist within projects requiring a range of functional specialisms. Many construction industry design-and-build contractors claim to use integrated project teams and Rolls-Royce claims to use integrated programme teams. However, it seems that the interpretation of integrated is different in both of these instances.

Looking at the use of IPTs within Rolls-Royce, the emphasis is upon the benefits of satisfying customers while also achieving the business goals of the organisation. A fundamental aspect of the philosophy behind IPTs as far as Rolls-Royce is concerned is that they go beyond the structure of the parent organisation, in that they are concerned with the processes within a project (or programme in the Rolls-Royce context). These processes are claimed to benefit from having the people working on them operating in the spirit of teamwork throughout the duration of the project. With this in mind, the key requirement of IPTs can be argued to be that they should be set up right at the beginning of the project (remember that the project life-cycle has at least three stages) as this is when the team members have the greatest opportunity to positively affect project costs. A commonly held assertion in design circles is that the first 20–30% of a product's design development locks in 80–70% of the project cost. However, while IPTs are intended to exist throughout the project, membership of them can be on either a full-time or a part-time basis.

Members of IPTs can be customers, product/component suppliers and project partners. In each case the important consideration is that membership is actually decided on the basis of the skills offered by a potential member matching up with the requirements of a project. IPTs therefore become the logical basis and focal point for measurement of performance achieved during the project's phases, and hence the appraisal and reward of their members. In order to achieve this there is the requirement to emphasise the issues of accountability, responsibility and authority within the team's role, which may vary during the project's life.

Furthermore, there is the need to address the behaviours that potential team members may exhibit. Such behaviours become as important as the traditional factors of technical and business skills which are normally sought in projects. This places particular responsibility upon the team leader, and it is suggested that leaders of IPTs should undergo specific training in this area. Large, longer-duration projects in particular should actively seek to put in place arrangements for the training, facilitating and coaching of team members as part of the need to continuously improve performance. Evidence of such training achieving an effective IPT can be found when:

- the IPT controls its environment;
- all required work disciplines are represented;
- all members are trained in team skills;
- the IPT leader has responsibility and authority with regard to all required resources; and
- the IPT has a clearly defined product to deliver, within a clearly defined scope, and start and finish points for the task.

Overall, the evidence should indicate that the IPT members have developed allegiance to the project rather than to their parent department or division. In this way, the IPT can be judged to have become self-sufficient for the duration of its task.

3.7.1 IPTs and project organisation structure

A point of interest with regard to the place organisation structures discussed previously is that, even though IPT membership may be on a part-time basis, Rolls-Royce views co-location of all team members as being preferable. This is particularly the case with regard to core members of the IPT. Overall, it is claimed that the costs of disruption in moving people to one site are more than offset by benefits in respect of concurrent engineering, unity of purpose and so on, irrespective of the functional specialisms involved.

It is also acknowledged that large projects may require a hierarchy of IPTs composed of three levels. The core team at the top of the hierarchy, whose constitution will change through the project phases, will provide team leaders for the IPTs on lower levels of the hierarchy. Comparison of this suggested project organisation structure with structures discussed previously indicates that, in common with the majority of project structures, it does not exactly match any of the standard models (more exotic, non-standard models will be discussed in Chapters 6 and 7). There are, for example, elements of the matrix structure, particularly the secondment model, about it, but the claims made for IPTs suggest that they overcome some of the disadvantages stated for the matrix type of structure. The important point is that project structures do not have to match exactly with any existing model – there is no law preventing project managers developing their own, bespoke structures (although the parent organisation may have a few policies and procedures covering the subject!). One example of how accepted thinking on organisation structures can change is the concept of virtual teams and these will be discussed further in the next chapter.

3.8 Conclusions

This chapter has covered a number of areas, each of which individually can be expanded to generate a much larger discussion. However, the intention here has been to provide an introduction that, while it does not intimidate those who are new to the area, allows some preparation for more advanced study later in the book. At this point it is important that you are reasonably comfortable with your understanding of such issues as the differences, advantages and disadvantages of the three basic organisation structures of functional, pure project and matrix. The consideration of the other structures covered in this chapter, such as place structures, is also useful in widening your horizons with regard to further possible organisation structures.

The discussion of integrated project teams is relevant in that it provides preparation with regard to the possibilities for organising the resources, particularly the human resources, required by projects in innovative ways. Always part of such a discussion are the potential problems, along with the potential benefits, of such approaches. Realism in managing projects is just as important as humour!

4 FURTHER FACTORS IN A POSSIBLE MODEL FOR ORGANISATION STRUCTURE DESIGN

De duobos malis, minus est semper eligendum – of two evils, the lesser is always to be chosen.

Introduction

The previous chapters have introduced different perspectives on the process of designing project organisation structures. A constant factor within the process has been identified as the need to respond to change. In times gone by (when life was so much simpler!), projects needed to consider change in only two forms. The first was the slow rate of change in the project's external environment and, even on a project of long duration, this may have been so slow as to be negligible. The second form was that which today may be referred to in reassuringly scientific terms, such as unexpected change events, or more traditionally fatalistic terms, such as accidents, but which would previously have generally been regarded as being in the category of acts of God.

In the more recent past the situation has become one where project teams have increasingly recognised that many of the events (not in the project planning context of start and finish events for activities) that happen during a project's lifetime are actually amenable to the imposition of control. This has fed a mentality that has sought to eliminate risk by finding ways of exerting control over more and more of the events within a project. This has had the benefit of increasing knowledge about, and understanding of, the nature of projects and this chapter looks at a number of factors that may be involved in this. Some of these factors have been introduced previously, in which case further information about them will be added. Other factors are introduced for the first time, in which case they should be considered

112

in the overall context of establishing how a project may be structured in response to change.

4.1 *Changes in the external environment*

The environment external to a project will inevitably change. This fact need not be a problem of itself, as the rate of change may be so slow as to be barely noticeable over the duration of the project. This seems to have been the case in many pre-industrial projects where, while it was possible to complete the intended change (construction of a castle or similar) at a relatively rapid rate, that change was internal to the project. During such a rapid period of internal change, the external environment was typically slow to change. Even so, change was soon recognised as an important factor; Heraclitus, the ancient scholar, noted that 'nothing is permanent but change'. However, it was not always the case that external change occurred slowly – the occasional border dispute, insurrection or change of head of state could bring a quite sudden change, such as a decision to terminate a particular project. The situation facing the majority of contemporary projects has generally involved a more unstable external environment that has presented the project team with greater levels of uncertainty. The tendency has been to respond to this environmental uncertainty by seeking to impose control.

At a basic level the imposed control has been in the form of various scientific management techniques, but these are essentially concerned with modulating the project's internal environment so as to minimise the impact of external change. A more ambitious form of control can be found in relation to those projects taking place in a centrally planned economy. Such economies, which are fairly scarce at present, refute the demand-driven free-market approach of the more capitalist economies and seek to direct through central planning the rate, type and extent of change that takes place. However, even in the free market economies, governments frequently seek to control change through the funding of large infrastructure projects, so it is debatable just how free such economies actually are. Nonetheless, it is arguable that centrally planned economies present fewer change events to a project than is the case with free market economies. Possibly the largest infrastructure project to be found in what may be classed as a centrally planned economy at present is the Yangtze Three Gorges Project (YTGP) in China. This project was introduced as a case study in Chapter 3 and will be discussed further here in the context of external environment change.

4.1.1 Further discussion of Case study 2

YTGP was supported by a longer than average lead-in and feasibility period, with a study which commenced in the late 1950s and ran through until 1992 when the National People's Congress passed a resolution to commence construction of the project. A feasibility study lasting around 40 years would normally be expected to have identified and quantified the risks and problems faced by such a large and expensive project. The study was classed by the YTGP Development Corporation as being exceptional in terms of its duration, scale and the number of organisations and individuals involved (CYTGPDC 1999). This study comprised, amongst other factors, the relationship between the project and the planning for the overall harnessing and further development of the Yangtze valley; cost-benefit analysis of the benefits versus scale of the project; the structural layout of the project including type and style of the main structures; air defence safety (after all, a September 11 style attack on the completed dam could result in a massive disaster); and planning for the power system. Perhaps perversely, the project seems to be at most risk from an external factor whose delivery cannot be planned and controlled by any government: earthquakes.

A project of YTGP's size would normally be seen to be at most risk from external environment changes such as capping or withdrawal of funding (shades of the Channel Tunnel!) and change of government. As far as the issue of funding is concerned, the project was set up on the basis of socialist market economy and international practice principles. The development corporation is an autonomous economic entity and ultimate owner and operator of the completed dam and is responsible for financing and reimbursement of investment in the project. This has been estimated to be in the region of 203.9 billion yuan by the completion of the project. However, it is expected that the project could become self-financing from 2003 as it begins to generate power, so the main problem has been to secure investment up to that point; a period of approximately 11 years. Investment has been raised through two mechanisms:

- A tax on power generated elsewhere throughout China. This is in the region of 0.3–0.7%/kWh. In 1997, China generated 1100 billion kWh and the tax had to that point covered approximately 50% of the construction costs. The revenue generated by an existing power plant at Gezhouba has also contributed to the required investment.

- Loans in the form of domestic loans from the China Development Bank and loans from foreign banks. Foreign loans seem to account for a relatively small proportion of the total investment and therefore do not appear to represent a significant risk to the project.

As to the factor of a change in government, there seems little chance of this happening. While China has undoubtedly opened it borders over recent years, there is not going to be any significant change driven by a popular democratic movement. Given the long period of research and development that the state has invested in the project, it seems to have achieved iconic status and is not at risk from any foreseeable change of policy during the remaining period of construction. However, the issue of it being in an earthquake zone is one indication of a potential instability in the environment that frequently represents some level of risk to projects. The project appears to have been fortunate even in this regard, as the seismic activity in a zone of 15 miles around the dam site is rated as small in intensity and low in frequency. (The project seems to have developed a humour of its own based around Chinese amusement with statements such as 'You are the dam tourists', 'This is the dam site', 'You are on the dam bus', and so on. The relevance of humour to project organisations is discussed in Chapter 7.) Not all projects are so fortunate in having a relatively stable external environment.

4.1.2 Unstable environments

It is important to put this issue into context: an unstable environment is not automatically the same as an uncertain environment. This is perhaps best illustrated by looking at what factors are typically taken to indicate an uncertain environment. Pieters and Young (2000) suggest that factors such as the rate of technological change, increase in new number of products required to remain competitive, extent of globalisation and difficulty of access to information (this point is discussed in further detail in the next chapter), which is taken to include obtaining feedback from the environment, are all examples that indicate the level of uncertainty in an environment. The traditional response to increasing uncertainty has been to attempt to impose control and thereby build some certainty back into the environment. Lawrence and Lorsch (1967) developed the proposal suggested in their earlier work that organisations responded to uncertainty by introducing complexity into their structure: greater specialisation and so on. Unfortunately, this also had a negative impact with regard to integration across the organisation as a whole. As a general rule, increased complexity in the form of greater differentiation results in the need to actively manage the integration of the specialised units within the organisation environment. Since the time of Lawrence and Lorsch's research (which commenced in the 1950s), external environments are generally accepted to have become increasingly uncertain. However, have they also become increasingly unstable?

Unstable environments can be regarded as those in which there is a lack of balance amongst the components, from which a crisis may emerge without warning. The need to introduce integration mechanisms into environments that have become complex in order to address uncertainty has already been discussed and on this basis it seems reasonable to assume that an environment will tend to become unstable only after it has first become uncertain. However, it is possible to envisage a situation where a project team's management and data collection skills are so poor that they do not realise their external environment is uncertain and are only aware that it is also unstable when one or more crises confronts them. Such a project team seems doomed to fail from the outset and is therefore not a particularly good illustration of the relative impacts of uncertainty and instability.

A key difference between the two states is that uncertainty is largely based on assessments of the external environment – the rate of technological change, the number of new products required and so on are all capable of being assessed and measured over time. An unstable environment, however, may not allow for this. The level of instability, in terms of the imbalance in the environment, may be imperceptibly small right up until the point when the crisis emerges. This suggestion is evidenced to some extent by the nature of fractals (usually discussed in the context of chaos theory), which are known to contain minute perturbations that can have sudden effects on the system as a whole. These perturbations are so small that they would not be noticed unless you were specifically expending resources on searching for them. Nonetheless, any crisis that results from them can appear to do so without warning and impact on the project so negatively as to be deemed catastrophic.

Lientz and Rea (1999) suggest that crisis can be identified on the basis of factors such as the current state of an issue, the rate at which the situation is decaying, the increase in impact on the project of the situation, and the length of time that the issue has been active. They also note that the project team, who should be the first people affected by a crisis, may be so used to the factors identified being in a critical state that they no longer regard there to be a crisis: it is simply a case of business as usual. Examination of the suggested causes of a crisis further illustrate that an unstable environment differs from an uncertain one. There may be an inherent but previously unnoticed technical flaw in the project. A further possibility is essentially political in nature: an antagonist to the project may seek to sabotage the project in order to secure its resources.

Perhaps most relevant in the context of this book is the suggestion that the organisation structure and roles inhibit members from dealing satisfactorily with any number of project issues that ultimately prove detrimental to the project. This is not to say that an unstable environment need automatically be deemed unfavourable to the project. It has been suggested that the most creative forms of project management emerge in projects operating in an environment that is on the edge between stable and unstable conditions, although this does suggest a need for highly skilled project team members. A further point to consider then is whether an uncertain and/or unstable external environment automatically becomes unfavourable to the project.

4.2 Environments becoming unfavourable

As an environment becomes more uncertain and/or unstable, it is suggested that it also tends to become unfavourable – both uncertainty and instability can produce some nasty surprises. Nonetheless, it would be unrealistic to view them as automatically producing an unfavourable environment. Such an assertion does, of course, require that a longer perspective on what is favourable or unfavourable sometimes needs to be taken. One of the projects discussed in Chapter 2 (Santa Maria del Fiore) provides an example of how an apparently unfavourable change in its external environment ultimately produced a beneficial output from the project in addition to the intended output.

The dome of Santa Maria del Fiore was bigger than any other dome built before it and raised many technological problems. However, one that had not been immediately apparent was how to raise heavy stones, along with other, lighter materials, to the unprecedented height of the dome. Initially it seems to have been assumed that the existing lifting technology would suffice. However, this assumption was based on the belief that Brunelleschi's solution to constructing the dome would use existing technologies such as wooden centring to support it. In the event, Brunelleschi came up with a radical new approach to dome building that did not require centring. The traditional equipment for lifting materials (which was largely unchanged from the time of the Roman architect Vitruvius) would now be faced with a clear lift considerably greater than anything previously encountered. At this point the Opera del Duomo decided to hold a public competition to find a suitable alternative. However, none of the many inputs from the external environment really addressed the problem head-on – it seems that the Florentines suffered an uncharacteristic lapse in confidence and creativity when faced with this particular problem – which appears to have angered Brunelleschi so much that he designed a lifting device himself (King 2000). The result was unlike anything previously encountered and was largely regarded as a wonder in itself.

Even though the inspiration for the device's design is unclear, its efficiency when operated was such that it was able to raise an average of 50 loads per day (around one every ten minutes). In total, it is estimated that during its lifetime the hoist transported in the order of 70 million pounds of materials – a considerable project even by modern standards. The hoist was also surprising in that it appears the specialist theoretical knowledge required for its design and construction was largely unavailable at the time (1420) and so it represented a significant advance in technology that had benefits for other projects during the Renaissance.

4.2.1 Project boundary control

Controlling a project environment's boundary is a concept that has already been introduced and that will be dealt with in more detail in Chapter 5. At this point the intention is only to consider the implications of unfavourable/unstable/uncertain environments for the process of boundary control. The most significant consideration in this regard is the establishment of where on the open–closed continuum the project requires to be. Most projects, as opposed to long-term organisations operating on the basis of functional management, are argued to lie somewhere towards the centre of the continuum. The implication of this as far as structure is concerned is that the boundary can be used to exert some control over the imports and exports of the project. This situation can be compared with one where an organisation (and possibly some projects) lies at the open end of the continuum. Such an organisation has little or no control over the imports and/or exports to its internal environment. Miller and Rice (1970) cite an interesting example of such an organisation when they discuss the operation of a hospital accident and emergency (A&E) unit.

The basis of an A&E unit's *modus operandi* is that anyone who feels they have a need for the service provided is free to enter the unit's internal environment. They may remain there until they are able to be moved either to the hospital's main facility or are sufficiently 'repaired' to export themselves from the unit's environment. The unit's boundary therefore imposes little or no control over the imports and only slightly more control over its exports. This situation can be compared with an organisation (or very few projects) that lies at or towards the closed end of the spectrum. Such an organisation can impose considerable control over its imports in as much as it will accept only certain specified ones. Because of its tightly controlled import boundary, the system will be capable of exporting only a small number of products, each of which can be identified in advance. The system can therefore structure itself with the realisation that only a limited number of narrowly defined exports are required of it.

Such structures tend towards the traditional hierarchical model with its rigid method of operation – its external environment has little uncertainty and therefore requires minimal complexity from the organisation structure. Miller and Rice refer to this variability of boundary control as being 'permeability' – highly permeable boundaries give little control and are required for open systems, highly impermeable boundaries give considerable control and are required for closed systems. This situation raises the possibility of problems being experienced if the true nature of a system moves along the continuum during its lifetime, but it retains its original structure and attendant boundary control philosophy. Maintenance of the organisation structure over its lifetime therefore becomes a factor to be considered. However, maintenance should not be approached from the perspective of constraining the structure by constantly seeking to return it to its original form. Rather, it should be approached on the basis of keeping the structure healthy as it goes through the inevitable process of evolution over time. Or, in the words of the French writer André Gide: 'Loyalty to the past stops us seeing that tomorrow's joy will only come if today makes way for it.'

4.2.2 Structure evolution

Up to this point the book has considered the changing nature of project organisation structures as being an essentially slow process happening over the duration of many projects. Such change has largely come about in order to respond to variations in the external environment. It is arguable that many of these external environment variations have in fact imposed themselves on the project structure and it is therefore unrealistic to regard project structure changes as coming about due to the structure actively seeking to optimise itself voluntarily. For-profit organisations being what they are, they tend to avoid the implementation of changes that they see as adding costs or external accountability. There is therefore generally a need for the external environment to impose changes that it regards as being important. Protection of the environment against pollution is one example of an area in which the external environment will usually have to intervene by imposing desirable levels of protection. Left to their own devices, the majority of organisations would not consider environment protection worth including in their structure.

Such a lack of consideration is illustrated by an article in the *Guardian* newspaper covering the dumping of vehicles in El Salvador by less than ethical exporters and importers. Because El Salvador had not regarded air pollution as a problem prior to 2002, the government had seen no need to impose emissions regulations on road vehicles. This provided an opportunity for dealers in other countries that had imposed emissions standards to legally (but not ethically) export vehicles failing their emissions tests to El Salvador. The resulting trade seems to have contributed to such high levels of air pollution that El Salvador's government is now taking action to protect the environment. Some people have even suggested that the countries from which the failed vehicles originated should compensate El Salvador as its citizens have been using their lungs to recycle the emissions!

Project organisations have traditionally responded to externally imposed requirements through the expedient of adding a suitable function (in terms of meeting that requirement) to their hierarchy. A contemporary example can be found in the addition of a planning supervisor function to UK construction project structures in order to meet safety requirements imposed from the external environment. As was noted previously, this apparently simple action actually adds further complexity to the organisation as it increases discontinuity (Miller and Rice identified three areas where significant change represents discontinuity: territory, technology and time – perhaps worthwhile just checking your understanding of this against Chapter 1) and thereby the number of interfaces within the structure. In order to manage this increased complexity there is a need to introduce an integration function. Looking again at the UK construction industry, a number of the larger organisations have introduced a specific interface management function whose purpose is solely integrative.

While this form of evolution has doubtless proved beneficial, it is not without its problems, two of which in particular are its self-rein-

forcing nature and that it is reactive rather than anticipatory. The difficulties caused by the self-reinforcing nature (essentially that if a little
differentiation is good, then more must be better, particularly with
regard to control) are discussed in more detail in the next chapter, and
the focus here is solely on the reaction problem. This problem has two
components to it. The first is the reaction to specific changes imposed
by the external environment, as already discussed. The second, and
perhaps more important, is the inability of traditional structures to
anticipate changes (positive and negative) in the environment instead
of measuring deviation from planned rates of change, and only to respond to negative changes after the event. An example of this can be
found in a function common to many industries: quality control.

In the following chapter there is a discussion of Toyota's early attempts at lean manufacturing. It was noted that during the changeo-
ver to the new working practices, employees became more proactive.
This was manifested in their beginning to consider where problems
might occur rather than simply waiting for them to actually occur
(and then getting paid overtime to fix the problem, as was typical at
Ford and other western manufacturers). While lean manufacturing
structures may be one way of increasing anticipation in a project,
there may well be other structures that are even better in this respect.
Examination of this possibility will be carried out in more detail in
Chapter 6, but for now it is worth considering one relatively recent
evolution of project structures as found in the implementation of
virtual teams.

4.3 Virtual teams

Virtual teams have evolved as a result of the increasing capabilities
of computer hardware and software. As with virtual reality, virtual
teams can be argued to exist only in cyberspace, irrespective of where
individual team members may be located. Such a situation presents
a number of difficulties when considering the possibility of implementing an IPT as a virtual team, for example. One such difficulty is
that geographical collocation is seen as being preferable for IPT members, but this simply cannot happen for truly virtual teams. The best
that a virtual team can hope for in this regard is that team members
may be located within computing facilities as close as possible to the
project 'site', as this at least offers the possibility that members may be
able to visit the site (and each other) from time to time. In this sense,
the IT infrastructure of the parent organisation is the main constraint
on virtual team location. It is arguable, however, that the issue of lo-

cation is of lesser importance than the achievement of virtual team members' commitment to project objectives, and this is an issue that can be improved through a consideration of project structure.

Virtual teams present particular difficulties with regard to members achieving commitment to the project objectives. This is a problem due to factors such as team members not being able to physically separate themselves from their parent department or division, resulting in a possible conflict between their perception of, and allegiance to, department objectives and project objectives. Under such circumstances the role of the team leader becomes especially important, and Rolls-Royce, as one example, suggests the use of facilitated team training, the focus being to build up shared values between IPT members. In the case of a virtual team this may well be all that ultimately holds the players together.

The facilitation role may become so crucial to achieving ongoing commitment to project objectives that a facilitator is recommended as a permanent team member (would you consider this to be equivalent to the integration function discussed previously?). The inclusion of such a functional specialist is not typical of current project structures, where the emphasis tends to be upon those specialisms that are classed as being explicitly productive. However, this is not to say that some projects would not benefit from the inclusion in their organisation structure of such a specialism. There is also the need to consider issues such as the recognition and exercising of authority within such a team.

4.3.1 Virtual teams and authority in matrix structures

Given that truly virtual teams are functioning in cyberspace, with the only possibility of regular face-to-face interaction being video-conferencing (successful use of which requires the users to develop further skills), the team leader may well find that their authority is difficult to assert. This may be particularly so in situations where the leader of a collocated IPT, for example, would not normally expect to experience any problems. Dealing with problems such as anarchy, as one example, can be very difficult when it is not possible to gather the team members together and provoke discussion of the problem.

A key factor here is the generally held view that teams achieve their best successes as they develop over the duration of their life-cycle and reach the synergy phase of that life-cycle. At that point they attain what has previously been referred to as teamthink: effective thinking (typically stated as 2+2=5) where the output of the team is greater than the sum of the outputs of individuals within the team. Such thinking is achieved through the effective management of dialogue internal to the team and the beliefs and assumptions held by it. Achieving such effective management through the use of e-mails, for example, is particularly difficult. Given that virtual teams may be regarded as being at the cutting edge of project organisation thinking, it may well be most appropriate to consider a form of structure for them that was originally developed for projects in one of the more high-tech industries: the matrix structure.

Matrix structures can be regarded as having two organisational elements – the bureaucratic and the non-bureaucratic (commonly referred to as 'organic') – and these have many opposite characteristics. The attraction of matrix structures is their ability to allow specialists to be grouped in functional departments from which differing projects, which may be running at the same time, can be resourced. Matrix structures are, for this reason, also referred to as being simultaneous structures and are generally regarded as being the perfect transitional structure between industrial and post-industrial patterns of thought (Banner & Gagne 1995). However, this form of structure also has one or two problems to consider.

Bureaucratic elements act as the source of people for the organic elements in matrix structures in that they are typically composed of functional departments (marketing, planning, finance, legal, etc.). Consequently, these elements are typically highly rigid, hierarchical, characterised by political in-fighting (with winners and losers), based upon division of labour into narrow specialisations, and highly formalised as they run almost entirely on the basis of positional authority. However, they can safely be left to get on with the day-to-day running of their parent organisations, but project organisations are a different consideration.

The organic elements (human resources) of a project organisation are composed of temporary project teams, the members of which are sourced from functional departments within the parent organisation, to which they will return upon completion of the project. While the organic elements have the characteristics of being flexible and fluid, with the role of leader moving as different project phases require different specialisms, using job-enrichment strategies (giving people whole jobs rather than specialist elements), and running on sapiential (knowledge-based) authority, they do have one significant problem: they require mature people. In order for the matrix structure to achieve successful projects, the people involved in it have to behave in a mature manner. Matrix structures rely upon individuals who are not fixated on playing politics in the manner that is generally encouraged within bureaucratic elements, and who can put selfish concerns to one side (they are, to a greater extent, what are referred to as mature individuals). This can be regarded as another form of commitment and without this, matrix projects invariably fail.

Further problems relate to the tendency towards interminable debate of minute aspects of the project, with the result that decisions are never reached as they are passed between the two bosses and then debated by groups (rather than being left to those with sapiential authority) and infinite layering. Infinite layering is an extreme situation in which power dynamics, rather than the logic of structure design, become the driving force within a project and matrices are developed within matrices that lie within other matrices, and so on. A difficult situation arises when the structure adopts a life of its own and the layering becomes uncontrollable. The failure of a matrix project usually relates to the tendency of this structure towards anarchy (a state of confusion) due to the lack of a clear, single leader. It is generally accepted that people work better when they are responsible to only one leader or manager, but in matrix structures they have two: one in the bureaucratic element and one in the organic element. This leads to ambiguity that can open a power vacuum, providing an opportunity for managers seeking to achieve maximum power to step in, the result of which is failure to achieve the balance of power that is essential for the matrix structure to work.

Considering the situation where a virtual team finds that it has an IPT organisation structure (with its matrix-like features) imposed upon it, the team leader may well be faced with a form of structure that is claimed to have the following disadvantages:

- anarchy – perception that as soon as something is working it is changed;
- groupitis – an individual will not make a decision without group approval;
- stifling – the time required to achieve group consensus stifles the imaginative flair of individuals and the group becomes a barrier to rapid progress; and
- cost – excessive cost overheads can result.

Unfortunately, these disadvantages are not the only problems likely to arise. It is possible that a power struggle will commence between the team leader and one or more of the departmental/divisional managers. Within matrix structures the balance of power between these two roles is always finely balanced and if doubt emerges as to who is in charge, the functional manager usually wins and the project suffers. Certainly, in such circumstances the role of the facilitator becomes crucial, but of equal importance are the negotiating skills of the project manager – without strong negotiating skills, the leader of a virtual team operating within a matrix-type structure can be regarded as being at the mercy of those stakeholders who are functional managers.

Understanding test 1

As a means of checking your progress in assimilating the material covered so far, a scenario has been thoughtfully provided for you to analyse! Consider your response to the following.

A colleague approaches you for advice. She has been given responsibility for a high-technology product design project. The project is a joint venture with three other organisations, two of which are based in England while the third is based in France. Her line manager recently attended a management training course and returned with a high level of enthusiasm for IPTs. He has since repeatedly expressed this enthusiasm for IPTs and has now suggested that she adopts this approach for her new project. However, she has never attempted a project of this nature before and has no experience of selecting appropriate organisation structures. You, however, have experience in this area, hence her request for advice.

With the aid of relevant assumptions (which you should make explicit) about the product to be designed, guide your colleague through the organisation structure selection process so as to conclude what the most appropriate form of structure would be for the project. Reasons for rejecting and accepting particular structures should be made explicit.

4.3.2 Structure and flexibility

Flexibility may be regarded as a concept of relevance to both IPTs and virtual teams in that it raises further issues concerning organisation structure. The concept of flexibility covers a number of working practices and in general has been an under-researched area. However, two specific practices that form part of this concept have been the subject of research: temporal flexibility and locational flexibility. The latter practice is relevant to this section in that the need to achieve locational flexibility can present an organisation with considerable environmental variations. While individuals may be well aware of (and possibly have experienced) temporal flexible working practices such as flexitime and job sharing, it is probable that the only locational flexible working practice of which they are aware is home/teleworking, and it is this practice which we will focus on initially.

A survey (Barford & Churchouse 2000) of 202 UK companies found that 31% reported making use of home/teleworking locational practices. In comparison, 45% reported the use of the temporal flexibility practice of job sharing, and it is generally accepted that locational flexibility is not as common a practice as temporal flexibility. Indeed, it seems that the surge towards home-based working predicted during the late 1980s and early 1990s has not occurred in general, but there are specific areas that are strong practitioners of this approach. It may not be surprising, for example, to find that 86% of those survey respondents who operate in the computing industry claim to have employees working from home some or all of the time. Likewise, the largest category of home-based employees is that of sales staff (41%) – such staff are generally regarded as requiring little in the way of infrastructure support (a phone and a kettle?) and tend to be motivated by a commission-based approach to performance management. However, even the British government employs people on a locationally flexible basis, with one example being access inspectors dealing with rights of way cases. These individuals are offered a 'home-office' package that allows them to set up a functioning office (telephone, fax, e-mail, internet, etc.) in a spare room of their home.

Given that locational flexibility is on the increase, the reasons cited by companies for introducing it are of interest. Perhaps the most important reason would be assumed to be a desire to reduce costs. Not so, apparently, as only 5% of companies cited that reason. Perhaps, then, because their customers felt it gave them a better service? Only 14% of companies cited customer demand as a reason. The largest single reason identified was employee demand (38%), followed closely by preparation for future business requirements at 32%. Perhaps the future business requirements related to an expectation by organisations that their employees would soon be demanding to work from home? Unfortunately, the detail of such requirements was not forthcoming, but it could be a useful exercise to consider for a few moments whether your organisation could be faced by a future business requirement that could encourage it to introduce locational flexibility.

One commentator on the survey results raised the interesting point that it was unlikely that one individual, particularly in a large organisation, would be able to respond accurately to all the questions raised by the survey (Flexibility 2000). It is entirely possible that your organisation is operating some form of locational flexibility (and/or temporal flexibility) of which you are unaware. Again, a useful exercise could be to make a few inquiries to determine whether such practices are operating within your organisation and how those involved view the reality of flexible location work schemes. The perception of those involved is almost always positive – the survey discussed here found that 82% of employers and 92% of employees believed their schemes to be successful. However, it is suggested that there is an underlying principle in that organisations promote flexibility only where they see it as an effective way of dealing with particular tasks by allowing work to be done at the most appropriate location. The question then becomes one of who decides on the most appropriate location or times. Such a question is perhaps relevant to non-traditional structures such as those found in transformational organisations.

4.4 Transformational organisation structures

In the discussion of virtual teams and matrix structures earlier in this chapter the concept of industrial and post-industrial forms of organisation was mentioned briefly. Post-industrial structures can be regarded as being essentially transformational, while industrial structures can be regarded as being transactional. Given that many countries within the developed world now view themselves as being post-industrial, it would seem reasonable to expect that organisations within those countries would show evidence of having adopted transformational structures. However, prior to looking for evidence of such a change, there is a need to consider what is meant when reference is made to transformational structures.

Banner and Gagne (1995) suggest that an important aspect of a transformational organisation structure is what may also be referred to as effectiveness. In this context, the transformational perspective on effectiveness considers the organisation as operating as one part of a seamless, organic whole that is larger than it. The key aspect within this perspective concerns itself with the requirement for the organisation to be operating seamlessly with its external environment. In essence, the organisation structure puts to one side the systems concept of the boundary: the truly transformational organisation is structured without a boundary.

Throughout the book there are various discussions of the different ways in which the boundary concept may be used by an organisation. In the majority of cases, the type of use is focused on attempts to control the environment that is 'out there' (external to the project) so as to ensure the success of the environment 'in here' (internal to the project). It may well be that some of these discussions have appeared radical. However, as a benchmark for radical thinking they are not as suitable as the suggestion that organisations do not just control their external environment; they actually create it. Maturana and Varela (1980) posited that an organisation only gives the appearance of adapting to its external environment. Acceptance of this appearance as being accurate is the action of, essentially, a transactional mindset that cannot envisage any other interpretation of the information that it gathers. There is, though, another interpretation – the organisation (as represented by a collective paradigm) projects itself outwards, views this projection as being its external environment and then reacts to it as if it were truly external. What is actually happening is that the organisation 'creates' an environment that is most like itself and therefore ensures its survival through self-replication. Such a suggestion could, admittedly, take a little while to get to grips with. In order to help in this process, the discussion will return to the issue of effectiveness.

In the Maturana and Varela model, the organisation's ability to sense what will be the most appropriate role to play within its externally projected self (the non-existent external environment) is the key measure of its effectiveness. This is in fact not far removed from the earlier discussion of energy release by project participants. Such participants are the collective paradigm representing the values and beliefs of the organisation (and the groups and individuals forming it). Their decisions with regard to what is relevant or irrelevant, important or unimportant, significant or insignificant, actually create the organisation's external environment. In a transactional organisation this environment would result from the collective paradigm of 'what is in this for me?'. The members of that paradigm could be viewed in D'Herbemont and Cesar's terms as being antagonists to the project organisation and therefore create an external environment that is unfavourable. Such an organisation would be regarded as having a low level of effectiveness: the environment selected by it (or at least by its collective paradigm) is not that most suited to its survival.

Taking the same model and replacing the collective paradigm with one based on a transformational organisation results in a more effective organisation. The paradigm now becomes one of asking 'what can we do to help achieve the objective of the whole?' and the resulting environment will be much more suitable to the organisation's continued survival. Such a paradigm should not be taken to suggest that every player is nothing more than a drone. Banner and Gagne in fact suggest that a transformational organisation's internal environment would support players in finding their true identity in the context of the organisation. This true identity would not be one constrained by a defined functional role as in a transactional organisation. Perhaps the most significant feature of such a collective paradigm for the issue of organisation structure is that it would allow the leadership role to move between the players in response to the emerging needs of the organisation. For a project organisation structure this would mean that the role of project manager would not be played by the same person throughout the project's duration. Indeed, the situation can be envisaged where there is no pre-planned order of succession for the project manager role – the collective paradigm would decide when the most opportune time to change was about to occur. It would also decide on who would be the most suitable of its members to play that role at that time. Leadership would no longer be related to position. In the new paradigm it would be recognition of an individual's sapiential authority.

4.4.1 Sapiential authority

We have already introduced the concept of sapiential authority and will deal with it here in more detail due to its significance within the study of transformational organisation structures. Sapiential authority is awarded in recognition of the changes in organisation environments whereby workers in some organisations have grown to be more knowledgeable than their leaders. This seems to occur particularly in those industries where knowledge is both embedded in the individual and changes rapidly over time. Such industries have seen the emergence of what Banner and Gagne (1995) refer to as knowledge workers, and it is this type of worker that has brought about the concept of sapiential authority (a term coined by Robert Theobald 1970). Because an individual may have a greater level of knowledge than his or her leader, who may have a formal or positional authority within a transactional organisation, the question of obedience is raised. There seems to be a decreasing level of willingness in the developed nations generally to simply obey an instruction because it comes from someone with a greater positional authority and this may be a factor that has allowed sapiential authority to emerge as an issue to be considered when structuring an organisation.

An important factor with regard to sapiential authority and project organisation structure is that this type of authority is strongly linked to organisation culture. Most of us will have experienced at some time the pressure that can be exerted by an organisation to adopt an identity based upon working for that organisation. We may see ourselves as a 'BMW' individual or an 'AMEC' individual and so on. The organisation develops a set of values of beliefs that 'its' people are expected to uphold (the aforementioned collective paradigm). However, if an organisation is to survive it has to be willing to change some of those values and beliefs from time to time. Within a transformational organisation the values and beliefs seem to be less rigid in that the intention is one of the collective paradigm driving the organisation rather than the organisation driving the collective paradigm. The emphasis therefore becomes one of valuing the knowledge and experience of individual players. On this basis sapiential authority comes in two forms: technical knowledge based and personal maturity based.

Technical knowledge as a basis of measuring an individual's value to an organisation is a well-established concept within the transactional paradigm. Individuals are usually recruited on the basis of their ability to provide a type and/or level of technical knowledge that a particular organisation requires. If the individual is sufficiently knowledgeable, they will be given a higher level of formal authority; they then trade on the status of this higher level in order to secure obedience. In this situation it is the position that carries the authority, not the individual. In moving the goalposts so that the individual carries the authority, irrespective of their position within the organisation, their title ceases to have any reference to anything other than describing their job. The organisation structure must then become more flexible in order to ensure that it can continue to function on the basis of authority moving frequently between individual players. Whichever player is seen to have the most relevant technical knowledge to the situation facing the project at a given time will become the project leader. When the situation changes and different technical knowledge is required, a new leader will be identified. An important point to note here, and one that will be returned to, is that the term is 'leader' rather than 'manager'. In transformational organisations one objective is for all the players to become self-managing.

Discussion of the concept of self-management presents an opportunity to consider a particular type of self-manager whom it is possible to confuse with the type intended within transformational organisations: the Dionysus manager. Handy (1999) introduces the concept of Dionysus in the context of self-oriented individuals who come together in what he regards as possibly the most minimally structured organisation, for which he coins the term 'cluster'. The cluster organisation has a strong person orientation and Handy cites examples such as hippy communes, architects' partnerships, barristers' chambers and so on. All of these are claimed to have the common feature that they exist only to serve the needs of the people within them (rather than the more usual case of the individuals within serving the needs of the organisation) and have no overarching objectives. Consequently, it can be difficult to exert any external control over them as the individuals forming them have only minimal interest at best in the success of the organisation as an entity in its own right.

Handy suggests that organisations having a predominant person culture are rare, but that the individuals who would be most suited to such environments are not. The characteristics of such individuals could be mistakenly taken as the basis of sapiential authority in that they have strong technical or specialist knowledge. However, the crucial aspect of their nature is that they do not ask what they can do to help achieve the objective of the whole (as in a true transformational organisation). The emphasis of the Dionysus manager is on what good the organisation can do for him or her, and as a consequence they are difficult to manage, particularly within a transactional structure. Handy illustrates this point by noting that none of the traditional transactional forms of power (positional, resource, expert, coercive and personal) particularly concerns such individuals. In short, they can validly be regarded as having good technical knowledge, but they should not be regarded as having, or responding to, sapiential authority.

While dealing with the issue of technical knowledge, it may be worth trying a little exercise. Appendix 1 contains a vast number of project management terms and their definitions. Try out your own level of technical (in terms of the management of projects) knowledge by selecting at random 20 terms. Do not read the definition! Try to produce your own definition and then check it against the one supplied. There is no pass or fail on this one, but if your success rate is fairly low, you might want to consider the implications for your project management sapiential authority. However, even if your score is low (or even very low), all is not lost – there is always the possibility of achieving sapiential authority based on personal maturity. This type of authority is sometimes also referred to as spiritual authority (Banner & Gagne, 1995) but should not be mistaken for authority based on depth of religious belief. The emphasis is actually upon the setting of a good example to others. Players who have consistently exhibited integrity will be recognised as people who can be trusted to give honest advice and be supportive of creative processes. Perhaps this more closely fits what you would see as your true identity, the role that you are called upon to adopt most frequently. If not, it may be possible to improve your sapiential authority based upon technical knowledge through the use of aids such as the various bodies of knowledge that have been developed.

4.4.2 Use of bodies of knowledge

Bodies of knowledge can be regarded as a further development of the standards-based recruitment mechanisms typical of transactional organisations. As a society evolves from pre-industrial to industrial and then to post-industrial, the basis on which individuals are selected to join a particular organisation tends to go through a standard model of change. This can be evidenced by examination of what may be referred to as recruitment (in the widest possible meaning of the word) standards for an organisation and the formation of boundaries by professional bodies.

Belbin (1993) considered recruitment standards in the context of three historical periods: pre-industrial, industrial and post-industrial. He then evaluated the type of criteria used by organisations for the assigning of work in each of these periods, the suggestion being that they were by category (age, gender, etc.) in the pre-industrial period, by qualification (skills, experience, etc.) in the industrial period, and by what he referred to as 'person shape' in the post-industrial period. Having established the criteria for assigning work, the methods of selection (recruitment) for each period were then identified as being visual inspection (pre-industrial), certificates and/or selection panels (industrial), and computer matching and/or counselling interview (post-industrial). This profile suggests a varying approach to the recognition of knowledge by organisations over time and that they need to be aware of any implications of such variation for the effectiveness of their project structures.

The situation with regard to professionals and professional bodies (collectively regarded as professionalism) is slightly different in that they are argued to develop over time in a more closed system manner. Barrington (1960) identified six stages in the development of professionalism, commencing with the foundation of a voluntary association having an objective to exclude anyone likely to lower public esteem of the association and the profession it represents. This is followed firstly by the development of an explicit code of conduct and secondly by the imposition of a series of tests. Having established an educational framework, the association then typically proceeds to exert control over any relevant educational institutions through the use of accreditation requirements and so on. This then gives the association a basis for the move from the national to the international level before proceeding to the final stage of achieving statutory regulation. At that point, project structures can find it increasingly difficult to recruit non-association members of a particular profession. One obvious implication of this situation is that resources (individual professionals) may become difficult to obtain, particularly if the association adopts a highly closed system having the effect of keeping the numbers of qualified professionals at a low level.

The industrial period of the selection model uses as its basis criteria that are essentially a means of differentiation between specific functional specialisms. As complexity is added to the organisation structure, organisations need to be able to pigeon-hole individuals with regard to separate specialisms. Bodies of knowledge make a contribution towards this process by identifying what knowledge is relevant (and by implication what knowledge is irrelevant) and bringing it together in a tidy bundle that can be taken to represent a particular role. As a means of differentiation (on the basis of knowledge as an example of territory), bodies of knowledge can therefore be divisive and are certainly a factor in the culture of a professional body or institute. The Association for Project Management body of knowledge is a good example of how knowledge tends to be categorised into discrete bundles. The fourth edition of the body can be accessed through the APM's website (www.apm.org.uk). The body suggests six knowledge areas:

- strategic (value, risk, and quality management, etc.);
- control (resource, budgeting, cost management, plus change control);
- technical (requirement, technology, configuration management, etc.);
- commercial (financial management, procurement, legal awareness, etc.);
- organisational (opportunity, implementation, organisation structure, etc.); and
- people (teamwork, leadership, conflict management, etc.).

It is interesting that the issue of organisation structure covers approximately two-thirds of a page and identifies only the three basic structures of function, project and matrix. The introduction of the organisational section does note, however, that the type of organisation should be an appropriate response to the key performance indicators and critical success factors for a specific project (APM 2000). These issues will be discussed further in Chapter 6.

This approach has the benefit of allowing a means of assessing the competences possessed by an individual, a process that may be as simple as ticking boxes on a list of competences. One of the problems with this basic approach is that it does not really give much indication of the level of competence that an individual can achieve. Sapiential authority based on technical knowledge will be difficult to achieve if other members of the collective paradigm cannot assess the level and extent of an individual's knowledge. It is therefore debatable whether bodies of knowledge will have any significant use within the post-industrial transformational organisations that are starting to emerge.

A further consideration is that publications such as the APM body of knowledge are not actually true bodies of knowledge. For example, an important factor in the success of the planning function is the planner's knowledge. This is important in that there is a tendency for some organisations to regard information and knowledge as being the same, but there are subtle differences between information and knowledge. Knowledge can be regarded as being the range of a person's information, while information can be regarded as being individual items within a person's knowledge. Individual items of information truly become knowledge when they are combined. This concept was introduced in Chapter 1 and this could be an opportune moment to indulge in a little revision. The point is further reinforced by the previous suggestion that sapiential authority can be achieved through both technical knowledge and personal maturity, with the latter representing 'life' knowledge.

4.5 Conclusions

This chapter has moved further away from the transactional perspective introduced in earlier chapters and started to introduce some of the more radical concepts that provide the basis for future chapters. The most significant new concepts introduced here are those of transformational organisations and sapiential authority. An understanding of both concepts is important if the links between the industrial-period approach to structuring projects (and organisations generally) and the approaches that may be possible in the post-industrial period are to be established effectively. These links are, in essence, conduits to change, both of the nature of organisation structures and of the individual mindset. The problem of change, and in particular how the organisation structure seeks to respond to it (traditionally through attempts to constrain through imposing control), will be developed further in the next chapter.

5 CONTROL IS NOT TOTAL

Fugit irreparable tempus – time irretrievably is flying.

Introduction

The most significant legacy of the so-called scientific management approach (as espoused by Gilbreth, the originator of work study, and represented by concepts such as therbligs, or micro-movements) as far as the subject of this book is concerned is a mindset that believes almost fervently in the act of control. While it cannot be calculated how many projects have been planned and structured on the basis of control being the acme of project management, there can be little doubt that such projects are in the majority. This fixation on control has tended to result in organisation structures that have been too controlling, in that they have experienced difficulty in accommodating anything but planned change. Such a situation is fine up to a point. After all, projects are essentially all about achieving change: factories where there was previously none; more powerful aircraft; new and better pharmaceuticals; and so on. However, not all changes can be planned. The previous discussion of possible versus probable change illustrated one factor that supports this dictum. This is not to say that the control function is irrelevant – it is obviously of importance – but this chapter will examine some of the issues in the argument that control should be balanced by flexibility, both personal and structural.

The nature of the argument can be illustrated by returning to the issue of scientific management. Gilbreth developed the concept of the therblig (an anagram of Gilbreth, but that was probably already obvious) as a result of his desire to find more efficient ways of working at production tasks such as bricklaying. The therblig is essentially a small (micro) movement and Gilbreth identified 17 of them. By combining individual therbligs into groups, actions that would result in the efficient completion of a task could be planned. Once these actions were planned, the basis of a control standard became available and conformity of action became possible. There is no intention to suggest here that the resulting emphasis on work study was, or is in its more recent forms such as method-time-measurement (MTM), not a factor in significantly improving the efficiency of production processes. However, there are two factors to take into account when considering the value of such an approach. The first is that humans are not machines, and some of them actually like to have some control over their work themselves, so at what point does the desire for control by the planner overrule the desire for control by the worker? This could be argued to be a metaphor for the tension evident between transactional and transformational approaches to organisation structure design, but that will be left to Chapter 7.

The second factor concerns the ability to respond to unexpected change events. It would seem a ridiculous proposition that, in the event of a fire in the workplace, all workers should leave the area using only those actions that have been planned on the basis of being the most efficient combination of therbligs. In fact, individuals would be given the freedom to evacuate the area using the actions that they felt most appropriate, subject to a general guidance concerning not endangering others by their actions. They therefore have some degree of flexibility in selecting their response to a problem – they are trusted to make reasonable decisions and in the majority of such instances that is exactly what they do within the extent of their circumstances.

The issue of freedom of choice can be examined further by considering the concept of degrees of freedom. This can be found in the work-study discipline and relates to the number of different directions in which a human joint can move. Such a concept recognises that there is a physical constraint on the extent of movements available and that any attempt to exceed that constraint will result in damage to the organism concerned. There is obviously a safety consideration that suggests it makes sense to work within those constraints, but would there be any point in imposing artificial constraints that reduce the degrees of freedom available? Few suitable situations come to mind and they can perhaps be regarded as the exceptions that prove the rule (or heuristic at least!). So why do many organisations structure their projects in such a way as to artificially constrain the degrees of freedom available to the project team? Arguably, because they believe that control is the key to success.

5.1 Project boundaries

Project boundaries were introduced previously in the context of systems theories concerning internal and external environments. The emphasis given to boundaries was essentially one of control, insofar as they were seen as being a contributor to the achievement of an acceptable level of quality. It is certainly possible to view boundaries as being impassable barriers through which the project team exerts absolute control over the resources entering and leaving the project environment, although such a perspective is impossible to turn into actuality. However, this perspective does seem to be the basis on which many transactionally oriented organisations structure their projects, in that they convince themselves that they can truly achieve such a level of control. To all intents and purposes, such a belief can only be maintained on the basis of smoke and mirrors – they conspire to delude themselves.

A more realistic approach is to regard boundaries as being, to varying degrees, porous in nature. After all, it was previously acknowledged that the project team is generally able to exert only minimal influence, and near-negligible control, over the environment external to the project. It can, however, usually expect to impose greater control over the project's internal environment, albeit through the imposition of rigid, transactional structures. This would seem to suggest that the team should expect to be able to impose a level of control somewhere between these two extremes at their project's boundary. If the boundary is also regarded as being a meso environment, this suggested level of control would seem to be particularly relevant.

5.1.1 System boundary control

If, in fact, the project system boundary is regarded as being essentially a meso environment, it can be argued that the actual level of control imposed within that environment should be somewhere between the minimum and maximum levels that are possible. This is not to say that all resources (imports and exports) passing through the boundary should be treated with an equal level of control. It seems obvious that some resources (such as hazardous materials) will require a greater degree of control imposed on their transition through the boundary than others. In a sense, this is a similar argument to that used when considering the management of inventories. One quite basic approach to this management activity is known as the ABC method and involves the assessment of each resource that is purchased and held in an inventory by an organisation as being in category A, B or C.

A summary of the approach is that category A contains all of the most expensive resources (these usually represent around 10% of the total inventory) and these are given the most attention in terms of management input. B contains the medium-priced resources (around 20–30% of total inventory) and is given less attention. Category C contains everything else and receives only minimal attention. The argument is basically one of not spending more, in terms of paying for inventory management time, on a particular resource than it would cost to replace that resource if it was 'lost'. This is a variation of the suggestion that projects should emphasise the release of energy by their players – if the structure determines where energy should be released most appropriately, there is high probability of success. However, if the structure prevents this from happening, or causes energy to be released inappropriately, the probability of success is reduced. Further development of this suggestion raises the possibility of being able to regard non-human resources as simply being means to enable the release of energy by the human resources. If this suggestion is accepted, then all the quality control (QC), quality assurance (QA) and total quality management (TQM) hierarchies that the transactional approach to structuring project organisations have developed in order to impose control become less of a constraint on the project team members.

Such a suggestion can be balanced against the trend regarding inventory management in manufacturing industry production processes. Kalpakjian and Schmid (2001) note that the use of computer-aided process planning (CAPP) techniques, particularly those dealing with materials requirement planning (MRP), have evolved to require the keeping of full records of all inventories of materials, supplies, work in progress, orders, purchasing and scheduling. These systems are evolving further to include marketing and business activities, in which form they are referred to as enterprise resource planning (ERP). In the production context, Kalpakjian and Schmid claim a number of advantages for CAPP systems:

- Process plan standardisation reduces lead times and planning costs, and improves product consistency.
- Plans can be prepared for parts of similar shape or containing similar features.
- Plans can be modified to suit specific needs.
- Product routing sheets are prepared faster.
- Cost estimating and other functions can be incorporated.

At this point, consider whether these advantages would represent significant benefits for a typical project (rather than manufacturing processes) and if so, how CAPP technologies would fit into the project structure. For example, would CAPP be regarded as transactional or transformational? Is the rate of change on a typical project such that too great an emphasis on data collection (as required by MRP in particular) would act as a constraint on its flexibility? This suggestion can also be considered in the context of Heisenberg's Uncertainty Principle: the more accurately you attempt to measure the position of a particle, the less accurately you can measure its speed (Hawking 2001). While Heisenberg's principle is intended to function very much at the micro level, there is an interesting possibility that it may be valid at the meso level in project structures. The suggested form of this validity concerns the gathering of data: the more accurately you try to measure work completed (position on the programme, QA performance, etc.), the fewer the resources available to determine how the project is changing (its speed).

It is important to emphasise once again that the intention here is not to argue for the total removal of all QC, QA and TQM activities and hierarchies – there are many instances where these are in fact significant contributors to the success of a particular project. However, it is arguable that in many instances they are more appropriate to the application of functional management in longer-term production processes and that for possibly the majority of projects, they impose unnecessary constraints on certain activities. These activities may be individual or grouped, but in either case they are perhaps those that would, in traditional critical path network perspectives on project planning and management, be regarded as non-critical.

Unfortunately, there are instances where this traditional perspective misses the fact that an activity may well be identified by the maths as being non-critical and yet it is significant to achieving an appropriate release of energy within the project. After all, CPN (critical path network) techniques are concerned explicitly with time (and from there with cost), and any link with other measures of project success is frequently far from explicit. Identification of such links seems to rely largely on the expertise of the project team. This expertise can often be found to have filled in the gaps in the paper-based representation of the project when it is turned into the live project in the real world. A further blow to the generally perceived value of such quantitative approaches to structuring projects is that there is an increasing school of thought to the effect that they really contribute little to project success (Stacey 1992; Bolman & Deal 1995). An example can be found in the context of PERT.

You are probably aware that PERT was deemed to be the factor that allowed the successful completion of the US nuclear submarine programme around 40 years ago. These submarines were regarded as being hugely complex items containing many new technologies that had never been brought together before. PERT allowed, supposedly, all of the interfaces between these technologies to be identified and an optimum construction sequence planned out. The perceived success of PERT in the completion of the programme encouraged its use in further projects such as the Polaris missile development programme. Again, PERT was credited as being an important factor in the project's completion. However, there is now evidence emerging that suggests it was actually the people involved in the project who ensured its success, in that they were able to react to, and find solutions for, problems not identified by PERT. It seems that they formed themselves into something similar to Wenger's communities of practice (discussed in more detail later in this chapter) and thereby directed the release of energy themselves.

Of course, it could be argued that this is fine if the people involved truly have the best interest of the project at heart and they have the expertise to efficiently problem solve – precisely the sort of points that a transactionalist would use to argue that people need rules and guidance if they are to work efficiently! The problem is that this misses the point: scientific management (transactional approaches) was an important step in the evolution of management skills and knowledge, and was relevant to the environment in which it developed. However, that environment has changed and is continuing to change and the traditional response of dealing with problems by seeking to impose further control now merely adds to the problems that emerge unpredictably from an environment that could be regarded as being turbulent (of varying irregularity). This perspective on modern project environments as being turbulent is simply an extension of the earlier discussion of thermodynamics in that a transactional (scientific management) approach would regard the environment as being essentially laminar (clearly identifiable and separated layers of activity) and would seek to maintain it as such through the imposition of control and rigid organisation structures. Turbulent environments need a different approach.

The impact of the above points on the issue of project boundary control should be the realisation that the boundary should seek to control resources on the basis of their contribution to the release of energy within the project. On this basis, the structure of the project needs to achieve porous boundaries. They should be particularly porous with regard to information in that they allow the link between a resource and important, relevant information to be established. Unimportant and irrelevant information should be identified and reacted to by simply filing it in case it turns out to be important later. This link can be assessed in terms of the measure referred to as lambda.

5.1.2 Lambda and boundaries

The level of information flowing through an entity (a body, thing or being) has been suggested by studies in artificial life (ALife) as being vital to its survival. While it is not being suggested (yet) that an organisation represents life, artificial or otherwise, the lambda concept has a certain resonance with the manner in which organisations succeed or fail. An example of this point can be found later in this chapter when we will look at the issue of forms of contract as constructs of organisation structures. At this point it is worth examining further the suggestion that one common feature of organisations, whether at the national, multinational or project level, is that they need information. This point was introduced previously when discussing the input of resources to projects and it is worthwhile returning to it after the overview of the difficulties regarding acculturation. One non-standard perspective on this issue is that of lambda, but be warned: this section will cause those with more, shall we say, rigid thought processes a few difficulties.

The meaning of lambda used here is that which is used in the study of artificial life, in particular those entities that are known as cellular automata (CA). These exist within a virtual universe, versions of which can be run on the average home computer, and were first thought of a good few years ago as a possible means of carrying out universal computation. Since then they have been used to simulate a whole range of problems and suggest answers. They have also increasingly come to be regarded as alive, in that they process information – an ability which researchers are beginning to believe is the key to life's continued survival in the face of the universe's attempts to enforce entropy (Ward 1999). If the concept of a project as being a dynamic entity is returned to, and the above discussion of culture is added in, then the issue of information processing can be seen as being potentially of considerable importance to the survival of an organisation, particularly so in the context of project organisations.

Studies of CA universes have found that there are several different types of CA and each of these have different 'preferred' lambda values. So, for example, when lambda values in a CA universe are low (near 0), the universe produces little of interest and any emerging patterns of activity soon die out. However, as lambda levels increase, the universe becomes busier and different types of CA emerge. These CA produce stable patterns and rhythms and exist around a lambda level of 0.5. Once lambda levels move towards 1, the structures produced become unstable and quickly collapse. The most important aspect as far as this discussion is concerned is that of the CA behaviour around the 0.5 lambda level, as it is at this point that CAs begin to fall into two categories: complex and chaotic. Complex behaviour can be difficult to define – on what basis do we feel that we as humans are more complex than, say, chimps? After all, we share almost exactly the same DNA. When this question is applied to organisations, perhaps the answer becomes easier to establish, with the suggestion being that complexity relates to the extent of interfacing within an organisation and between the organisation and its external environment – the more interfacing involved, the more complex the organisation. Any consideration of multicultural projects, for example, suggests that they would be expected to reside towards the complex rather than simple end of the scale.

Chaotic behaviour is generally seen as being easier to define than is complex. There may be various references made to Chaos Theory, but mostly these are not valid as this theory does not quite fit with what most people actually mean when they describe something as being chaotic, the general inference being that the organisation or project is out of control. Perhaps fortunately, it is in fact possible to formally measure chaos and the extent to which an entity is behaving chaotically through measures such the Liapunov exponent (McCauley 1995). Under detailed examination, it can be shown that entities or systems that may have been thought, upon superficial examination, to be out of control may actually contain high levels of order. However, it is the region between complexity and chaos which is of the most interest in that what happens within this region is starting to be considered to be a vital factor in how systems deal with organising themselves, rather than being organised by others (such as project managers). This important region is referred to as a phase transition.

Research carried out on phase transitions has found that all systems, irrespective of whether they are tidal waves or businesses, pass through them at one or more points during their life-cycle and these points generally coincide with a critical stage in the system's development. The important consideration is that during such transitions order seems to emerge spontaneously – the system appears to organise itself without any outside assistance. Not only that, but the nature of the order which emerges seems to be consistent across a whole range of systems and it would therefore seem not to be too far-fetched to consider that the aforementioned tidal wave and business, at some point(s) in their life-cycle, behave in exactly the same way as their individual elements are overridden by the dynamics of information processing.

The trick for project managers would seem to be knowing that, irrespective of what organisation structure they impose on their project, at some point that structure will be overridden when the project develops a 'life' of its own as the information dynamic asserts itself. Knowing when this is going to happen and being prepared to use it would seem to be a very useful ability. However, phase transitions do not seem to be as clear-cut as may be the case with system boundaries, for example. Most project managers can identify boundaries within their projects: differences between activities, organisations (companies) involved in the project, and so on. These are quite traditional boundaries and project managers would not generally question their validity. However, a common feature of these boundaries is precisely the fact that they do seem clear-cut – bricklaying is not seen to be the same as carpentry, for example.

The problem when considering phase transitions is that they are in fact precisely that: transitory, ephemeral, or short-lived and consequently not consistent over time. A boundary between bricklaying and carpentry will remain consistent in the medium to long term as the nature of each trade will change only relatively slowly, and so a project manager may feel that such boundaries are reassuringly constant and can be relied upon. Unfortunately, phase transitions may happen slowly or quickly, and the point at which the impact of one phase is overtaken by that of the next phase cannot be determined with absolute certainty. Phase transitions are therefore likely to be considered to be worryingly uncertain as far as the majority of project managers are concerned. This may be particularly so for what have been referred to as knife-edge organisations.

Applying the second law of thermodynamics to the organisation of projects suggests that any attempt to provide order (in the form of a low-entropy system), so that production processes can operate without interruption, for example, is immediately faced with the universe's dislike of low-entropy systems. A significant consequence of this is that organisation structures are always seeking to balance the requirement for order to support the production processes and the natural tendency for such structures to collapse into disorder. Project management is perhaps particularly problematic in this regard and can therefore be viewed as operating on the knife-edge between order and disorder. This does not mean that any attempt to project manage is ultimately doomed to failure – the whole range of technologies with which we are surrounded on a daily basis proves that this is not the case. Nonetheless, there is the ever-present spectre of failure, and whole books have been written on the many spectacular failures that have occurred during the human race's recorded history. However, rather than concentrating on spectacular failure, it will be more profitable to concentrate on the fact that it would now seem possible to argue that such organisations spend most of their life-cycle moving in and out of phase transitions – a situation which is apparently shared with truly living organisms. Now there's something to think about.

5.2 Control of the human resource

Human resources can be controlled within organisations in two ways: psychological and legal. This may seem a simplistic statement – where does culture, at the national, regional or project levels, figure in it? Well, national culture can be argued to be a factor in the development of a nation's legal framework and is a contributor to the fact that there are different legal frameworks as you go around the world. Some nations enshrine within their legal framework the right of the individual to do all manner of things that may seem at least strange, and at worst abhorrent, to other nations. The national framework may then be modified at the regional level by a more relaxed or more rigorous interpretation, or by the addition of further responsibilities or specific exemptions. Similar arguments can be made for further factors in the behaviour of the human resource, such as politics and religion.

Difficulties such as these are an example of why the social sciences seems to prefer the modelling and prediction of behaviour at the level of the individual – the more individuals you bring together, the more difficult it is to predict how they are going to behave as a group or organisation. As individuals are brought together to form groups, teams or complete organisations, their behaviour tends to be modified by those around them, particularly if there are any charismatic leaders among them. The relevance of such issues to the action of controlling the human resource within a project organisation will be discussed in the following sections and should also be regarded as a contribution to the later chapters covering possible future structures (Chapters 6 and 7). A starting point in this discussion is the concept of IGOr.

5.2.1 The relevance of IGOr

An important point to clarify before proceeding further is that IGOr, in this context, does not refer to Dr Frankenstein's assistant (the role of humour in project organisations will be covered in Chapter 7, you will doubtless be glad to be informed). IGOr is one of those (near) acronyms that can lodge in the memory rather more easily than its individual components of individual, group and organisation would if listed without it. While it may seem somewhat flippant, this acronym (possibly more accurate to refer to it as a retrenchment, as a true acronym would be IGO and that is not nearly so amusing!) does relate to an important argument within organisations: they come about only where individuals acting alone cannot achieve their goals. Buchanan and Huczynski (1985), for example, assert that organisations do not have goals. This is despite the fact that many organisations have imposing mission statements that claim to encapsulate their goals. Whilst it is true that mission statements express goals, they are in fact either the goals agreed by the majority of the people forming that organisation or the goals imposed by a minority who control it. In the latter example it is not inconceivable that conflict within the organisation can occur.

While Buchanan and Huczynski appear to have been directing their comments primarily at production organisations, it would seem reasonable to regard them as also applying to project organisations. There does, however, seem to be an assumption by many project organisations that their goals are clear, particularly where a transactional management style is being used – the project needs to finish on time and within cost and quality limits. However, this assumption is not always valid, and even if it is, there is a need to consider that these goals are not sufficiently informative.

Tichy (1983) expresses the argument that effective organisations are identifiable by the presence of good strategic alignment: organisational components are aligned with each other, along with the political, cultural and technical systems also being aligned with each other. In order to achieve its goals an organisation must be able to evidence a clear understanding of them (we have already discussed the problem of mirage projects) and be able to bring together in a coherent structure its diverse elements. This coherent structure was suggested by Tichy as comprising parts of a chain as opposed to a series of independent links. Within this structure, emphasis needs to be placed on identifying and then strengthening the weakest link in the chain. One way of making that link easier to identify involves the reduction of the number of links in the chain and this is essentially the philosophy of the lean manufacturing (or production) concept. This concept gained ground in manufacturing industries such as the automotive industry. Womack *et al.* (1990) suggest that the origins of lean manufacturing can be traced back through the Toyota production system of the 1950s to the development work carried out by a production engineer (Taiichi Ohno) in the late 1940s, a time when western industries were largely sticking to the transactional organisation structures to achieve mass production. Such mass production systems are consistently identified as being characterised by the constraining of knowledge development and effort on the part of operatives.

Ohno concluded that his relatively small company (at one stage it had taken the company 13 years to produce 2685 cars at a time when one of Ford's Detroit factories was producing 7000 cars per day) could not compete by simply replicating the transactional approaches of its larger, western competitors. His strategy was to adopt what may be referred to as role versatility amongst the Toyota workforce – the opposite of the functional specialism espoused by scientific management. The impact of this emphasis on versatility was particularly evident in the body stamping plant. The traditional approach involved the use of hugely expensive dies that required absolute precision when being set up, thereby spawning a specialist role to reset them every time they needed overhauling or changing to produce a different body part. Ohno simplified the process to the point where the die-change time was reduced from one day to three minutes and was carried out by production line workers rather than by specialists. This removed one link from the chain (in effect, a reintroduction of integration rather than the traditional route of dealing with external environment changes by adding complexity through differentiation) and had the further benefit of making it cheaper per item to produce small batches than could be achieved by mass production large batches.

This cost reduction arose from two characteristics of the new approach: inventory costs were reduced (inventory for no more than three hours production was typical), and by incorporating the product of short runs almost immediately into the main production line processes, problems of quality (such as poor fitting parts) showed up more quickly and only small batches of parts would need to be scrapped ((Womack *et al.* 1990). However, the main benefit of the new system was that it made the operatives aware of the need to achieve consistent quality, and for this to happen a highly skilled and motivated workforce was required. At this point, the concept of the company as a community began to emerge.

5.2.2 Communities (of practice)

The Toyota community emerged from negotiations between the company management and unions in the late 1940s. Put simply, management was running out of credit and proposed to cut costs by sacking 25% of the workforce – not a popular solution. Eventually a deal was agreed whereby 25% of the workforce was sacked but the remainder were given two guarantees: lifetime employment and pay graded by seniority rather than job function, with bonuses linked to profitability. In order to sweeten the pill a little, the company president accepted responsibility for having to sack a quarter of his workforce and resigned. It then became apparent that the company had a new organisation structure: people would now be rewarded for their experience and knowledge rather than simple production ability. These same people also realised that it was now in their interest to be proactive and address problems before they happened rather than simply responding to them after they had happened. The overall effect was that the employees were now more valuable than machinery (old machinery could just be scrapped, whereas 'old' operatives could not, in fact it became essential to constantly upgrade the knowledge of the workforce) and so came to be regarded as part of the Toyota company community, where they were valued for their knowledge more than their transactional brethren's ability to move masses of metal down the production line.

This concept of company as community could perhaps be argued to be a further development of the enlightened, paternalistic employment practices of a small number of employers in Britain during the 18th and 19th centuries. Communities such as Saltaire, near Bradford, grew up around the company and provided education and decent living conditions for their employees. However, it is debatable whether even these communities regarded the employee as being valuable in the same manner as Toyota's community did. It was perhaps more a case of the owner realising his social responsibilities as a 'good' employer. Nonetheless, such communities can be regarded as one point of historical reference with regard to how communities could be structured around the production requirements of a company. A further point of reference is the work by Wenger on communities of practice.

Wenger (1998) suggests that communities of practice pervade our lives: they are everywhere. Perhaps the most relevant examples to this book are those communities related to our working lives. We have all experienced the manner in which we organise ourselves and our co-workers in order to get the job done. This is said by Wenger to involve developing a sense of ourselves that we feel comfortable with (we all have a self-image, irrespective of whether it is realistic or not), meeting the needs of the job with regard to what our employer and clients want from us, and (I particularly like this one) having some fun along the way. As with the Polaris programme, what people actually do at work may be considerably different from what their job description indicates they do. A community develops in order to achieve what that community sees as needing to be done. This generally involves a relatively small number of people, even if the organisation is a large one employing thousands, as is evidenced by the decision by Gore Associates, a manufacturing organisation, that no 'unit' should include more than 200 people – go beyond this size and most people seem to lose the ability to form a community containing everyone.

This point seems to be an important one in that Wenger sees communities of practice arising over time as a result of those involved sharing the sustained pursuit of a particular endeavour (this has similarities with the concept of the psychological contract – see section 5.3.2). If this sharing cannot be achieved because of a large number of people being involved, then a community cannot emerge. Project organisation structure therefore would seem to need to consider how relatively small communities may be achieved within the project environment. This is not a particularly radical statement in that it has long been argued that teams work best when they contain around ten people (Maylor 1996). Such small communities of practice are on a different scale to the Toyota company community. They do, however, resonate with Wenger's assertion that the endeavour being pursued is defined by working with others who share the same conditions.

The issue of shared working conditions should perhaps not be interpreted too literally – it does not appear that Wenger is suggesting communities can be delineated by one group working in conditions where the temperature is slightly higher or lower than that experienced by another group. It is perhaps more realistic to consider the shared environment in system terms, whereby one community may be sharing the experience of completing a particular subsystem. A particular feature of these communities is their reliance on collective learning over time to provide 'unofficial' solutions to problems. This could be regarded as an example of creativity, a factor that organisation structure can impact strongly upon and is therefore worth further examination.

5.2.3 Groupthink, teamthink and creativity

At this point it is useful to consider how communities may relate to teams in that it may have been assumed that they are essentially the same thing. An important question is to ask: when is a team not a team? The simple answer is when it is a group. One of the problems with the transactional approach to the management of projects and the development of project organisations' structures is that those involved can experience difficulty in telling the difference between a group and a team. It is not uncommon for project managers to refer to teams without truly appreciating that they are actually dealing either with groups or with a mixture of groups and teams. Manz and Neck (1995) referred to the differences between the two in terms of 'teamthink' and 'groupthink'. Their characteristics have been summarised (Chapter 3) as:

- Groupthink – members trying to agree with each other; no adequate discussion of alternatives; social pressure against divergent views exerted; self-censorship practised; no consideration of failure; collective efforts to rationalise; defective decision making.
- Teamthink – members engage in effective synergistic thinking; encouragement of divergent views; awareness of limitations; recognition of members' uniqueness; recognise ethical and moral aspects of decisions.

In relatively simple working environments, it can be quite straightforward to recognise which of the two types of thinking is prevalent. However, in more complex environments, particularly where the achievement of project objectives may call for a near total network of interdependencies between contributors, recognition can be difficult. In such circumstances it can be worthwhile looking for evidence of Maylor's (1996) three criteria for the recognition of a team:

- Output of the whole is greater than the sum of the outputs of the individuals.
- A greater range of options will be considered than with a group.
- Decisions made are more consistently effective than those made by a group.

Teamthink is also of value because of its benefit when creativity is required, and the project organisation structure will generally be guided by one of two perspectives (Reiss 1993) on the management of teams. While these perspectives were not originally identified as such, they do nonetheless appear congruent with the transactional and transformational approaches. The first perspective states that science and technology can aid in controlling (that dreaded word again!) and motivating the team. As was noted when discussing virtual teams, there is no doubt that technologies can support the development of new forms of team working. However, this should not be confused with the use of technology to control and motivate a team, an approach that very much indicates a transactional mindset. The second perspective is that most people will control and motivate themselves if given the opportunity. This seems to be the case with communities of practice also, so in this sense the two entities have a similarity. However, the issue of creativity suggests that the two entities may also have an important difference.

Creativity is suggested as being important at a particular stage of a project (Maylor 1996). This stage is characterised by activities such as conceptual analysis, proposal justification and reaching agreement. These activities are typically included in the planning phase of a project's life-cycle. Creativity is therefore seen in this model as an activity in which a team engages for only one phase out of the project's life-cycle. It is also suggested that creativity is one of three categories that a team can structure itself for, the suggestion being that each category will require a different structure. The other categories are tactical and problem solving. Furthermore, each category is suggested as requiring different characteristics of team behaviour. As to whether this will require different people for each category (which will require the reconstitution of the team as it moves between categories) is unclear but may be dependent upon the role versatility of the individuals forming the initial team. For example, the creative category requirement for independent thinkers may appear to contradict the tactical category's need for members who have loyalty. In the event that the team is reconstituted for each category, this precludes a continuous community of practice throughout the project duration. Instead, there will be at least three sequential communities formed.

A further point of relevance is to note that the approach of categorising teams differentiates between creativity and problem solving, which introduces a problem regarding the earlier statement that the finding of 'unofficial' solutions could be regarded as a form of creativity. It also appears to introduce a constraint in the form of the freedom of a team to act only being allowed within the context of a particular category (and therefore project phase). Such a situation would have adverse implications with regard to the opportunities for the team to learn and thereby become a more valuable project resource. Again, the categorisation approach is indicative of a transactional mindset in that it imposes constraints (control) on the team members. The extent to which freedom to act is allowed to teams is therefore suggested as being a relevant issue for decisions regarding project organisation structures.

5.2.4 Freedom to act

The concept of being free to act within a project environment has an important implication: individuals will accept this freedom and be willing to implement it rather than choose to regard it as permission to maintain the *status quo*. Hutchins (2001), for example, identifies individuals who suffer from paradigm lock when confronted with the implementation of change that they see as being in some way threatening. There are also individuals who are able to implement change even if it means that they must also change (refer back to the discussion of Wenger's suggested development of self-identity in section 5.2.2). These individuals are claimed to possess three characteristics:

- Able to fully subordinate what they are doing (personal aspirations) in order to achieve a (project or organisation) goal.
- Prepared to take responsibility and accept accountability for their actions. Recognise that they are part of the problem as well as the solution.
- Prepared to give and respond to leadership. Able to accept suggestions from any level within the organisation.

It would seem that when such individuals are given the freedom to act there is a high probability of them proving to be successful. However, there are two factors that can constrain the degree of freedom that an individual can be offered, and each of them is related to the nature of contracts.

5.3 Caveat emptor

Those who are working from a contingency-oriented background will tend to think in terms of contracts as being legally binding agreements between two or more parties who have some common interest (generally making profit). However, this section will introduce other perspectives on the issue of what is a contract, suggesting forms of agreement between parties, who may or may not be members of one or more organisations, which may not have been previously considered. As a starting point, there can be argued to be effectively only two types of contract: legal and psychological. Both of these have implications for organisation structure, if only that they involve some degree of what may be referred to as 'buying-in' by those who are to be part of the projects affected by them. In the case of legal contracts,

151

the buying-in has to take place within the context of responsibilities and functions that have the potential to be determined in a court of law. In such environments the act of buying-in by those who will form the project team may well be less on the basis of personal motivation than it is on the basis of (in effect) coercion.

Psychological contracts can create a more complex environment than can be created by legal contracts alone. This is due to the various levels at which buying-in can take place within such contracts, combined with the reduced potential for the context of the buying-in to be determined in a court of law. Note that the potential is reduced rather than eliminated – some aspects of psychological contracts can be embedded in legal contracts and the two forms should not be regarded as being entirely mutually exclusive. These issues will be expanded further in the following sections.

5.3.1 Legal contracts

Within the so-called developed nations there tends to be a highly diverse range of legal contracts (generally referred to in UK industrial sectors as standard forms) to choose from. This is particularly the case in those individualistic societies, such as the UK, which place great emphasis on establishing the rights and responsibilities of individuals, groups and organisations. The UK, for example, has in the region of 50 000 practising legal professionals. In comparison, France has approximately 17 000 legal professionals. The important aspect of this situation is that the UK construction industry (as just one example industry) uses a significant number (40+) of different standard forms of contract – far too diverse to deal with individually here, even if such coverage was appropriate. On top of the construction industry standard forms, there are those used in other industries, such as engineering. Finally, there is also a plethora of one-off forms that are possible in any industry. Consequently, it is suggested that it is of greater relevance to consider legal contracts as having two subsets (for the purpose of considering how organisation structures may develop): contingency contracts and postcontingency contracts. Such an approach may initially not seem to be either valid or to have any relevance to the subject. However, perseverance generally results in the identification of a link.

Evidence for the validity of the two subsets approach can be found by comparing two forms of construction contract, one in the UK and one in Denmark. During the 1980s one of the most commonly used forms of contract in the UK construction industry was known as JCT

80 (various JCT forms of contract are still widely used), a comprehensive document running to over 100 pages and giving the impression of covering just about every conceivable event detrimental to a project's progress. Also, just in case there was an event not explicitly covered, there was the possibility of amending the form to make it even more specific. The JCT 80 form was very much a child of the contingency approach, which is perhaps not particularly surprising given the adversarial reputation of the UK construction industry at the time.

The Danish construction industry has, historically, taken a rather different perspective on contractual relationships, preferring the familial rather than confrontational type of relationship. Consequently, at the same time that the British were using JCT 80, the Danes used the AB 72 form of contract. This ran to about six pages, which seems particularly concise, especially given that the form could be used for civil engineering projects as well as 'standard' construction projects. The emphasis within AB 72 was very much on non-hierarchical relationships, as would be expected of postcontingency thinking. It is argued that such a form of contract is evidence of a more mature and responsible industry that is not entirely focused upon answering the question of 'what is in it for me?'.

Taking such an approach further results in the suggestion that the contingency subset of legal contracts can be argued as having essentially two purposes: to formalise power and authority, and to manage/control change. We have touched on the issue of authority several times, particularly with regard to the difference between positional (flowing solely from a position within a hierarchy) and sapiential (reflecting recognition of knowledge/expertise possessed rather than position held) authority. However, the issue of power has only briefly been discussed and it is worth a brief outline at this point.

Power, in its organisational form, is a broader concept than that of authority. A key aspect of this is that while authority can be achieved simply by an individual's position within a hierarchy, power over another individual can be achieved only through the existence of some form of dependency relationship and as such can be used outside of positional authority. Indeed, in some instances it is possible that it can be used outside of the organisation itself and in such circumstances is referred to as illegitimate power, although this should not be taken as automatically making it a negative force.

Research on the issue of power has identified several forms: reward (the power to offer enhancement), coercive (the power to withdraw rewards), expert (arising from expertise and skill and therefore similar in a sense to sapiential authority), referent (power resulting from belief in an individual or their ideas) and the more complex owner-ship power (arising from acting on behalf of shareholders) (Walker 1996). The majority of these forms are capable of being formalised within a standard form of contract.

We have also touched on the issue of managing or controlling change, particularly with regard to change being a desirable outcome of a project. However, the contingency approach particularly seeks to identify and control, or at least manage, all sources of change within a project with a sufficient level of probability of actually happening so as to ensure (ah, the arrogance of it all) that only the change that has been formally planned takes place. This is very much an example of the project manager(s) seeking to establish an artificial boundary between the project and the wider environment. The boundary is artificial due to being based on a belief that everything that is within the project can be controlled through the mechanism of a standard form, thereby reducing risk and uncertainty. As far as the organisation structure is concerned, this situation is rather a cyclic one in that uncertainty with regard to change in a project requires an open and flexible structure. However, a project that seeks to remove risk and uncertainty will tend to adopt a closed and rigid structure. Risk, for example, will generally be dealt with by making someone else responsible for its assessment and management (one of the benefits of positional authority). Therefore, the more rigid an organisation structure is, the more 'certain' a contingency-oriented project manager will typically feel about his or her project, particularly if he or she is an immature (this term is not used in any value-laden sense) individual.

This, then, brings us on to the issue of psychological contracts.

5.3.2 Psychological contracts

As the legal contract between an individual and an organisation states the rate of exchange between what the individual and the organisation respectively have to offer (skills and payment, for example), the psychological contract deals with the expectations of the individual and the organisation. Prior to dealing with the individual types of psychological contract, there are three important points to consider:

- Most individuals have more than one contract with more than one organisation, as few people seek to achieve all their expectations within one contract.

- Contracts that are not perceived identically by the parties involved will be a source of conflict. Typically, organisations will expect more from the contract than individuals will.
- Motivation levels become predictable only when contracts are identically perceived by all parties involved.

(summarised from Handy 1999)

In the event of an organisation seeking to maximise the benefits of a psychological contract with its members rather than enforce a legal contract with them, the structure of the organisation should be such that the optimum form of psychological contract can be achieved. This requires the project manager to consider the three forms of psychological contract:

- Coercive – usually found in prisons, coercive unions and custodial mental hospitals, although it has been suggested that some schools, hospitals and factories have been found to use this form. The emphasis is upon rule and punishment – if an individual complies with the wishes of a powerful minority he or she will avoid punishment. Individuality is suppressed and conformity is emphasised (arguably, this is typical of groupthink situations).
- Calculative – a voluntary contract with an expressed rate of exchange between 'desired things' (money, promotion, work itself) that can be supplied by the organisation and services that can be supplied by the individual. This is currently the dominant form of psychological contract within industrial organisations for the majority of players. Care has to be taken to ensure that changes in the rate of exchange do not cause the player to regard the contract as becoming coercive.
- Co-operative – individuals tend to identify with the goals of the organisation and become creative in seeking to achieve them. Individuals are rewarded justly and are also given more freedom in selecting methods for achievement of goals. Day-to-day control is largely relinquished by the management team, but it retains overall control through the allocation of financial resources, etc. Co-operative contracts are becoming increasingly popular (possibly as a result of organisations moving to postcontingency approaches?) and are frequently described as 'empowerment' of players. However, project managers should be aware that players cannot be empowered if they do not want to be (consider the issues within the development of teamthink) – in such cases the contract then becomes coercive in nature.

The situation may appear to become further complicated when it is considered that there will almost certainly be more than one form of psychological contract operating in an organisation at any one time and that some players will be keen to move between different forms of contract as they develop and move through life changes. For example, novice players may be quite happy with a calculative contract while they are gaining in experience but then may wish to move to a co-operative contract when they have developed confidence in their expertise. Contracts may change between different functions within an organisation also, even if the individual moving between those functions has not changed. This can be apparent in the consideration of communities of practice, as discussed previously, in which group culture may be apparent only to those who are part of a particular 'community'. The nature of the psychological contract is therefore essentially invisible to anyone who is not a member of a particular community.

If the organisation is structured so as to enforce only one type of contract, such as co-operative contracts, for example, its membership will be largely (if not wholly) constituted of unhappy players. However, it should never be assumed that every member of such an organisation is automatically unhappy with such a situation – people can be very strange animals at times.

Memory test 5

Before working through the following sections, try to answer these questions:

(1) What are the two basic types of contract?
(2) How is the concept of power a broader one than that of authority?
(3) Outline three of the five types of power identified.
(4) What are the three forms of psychological contract?

5.3.3 Contracts and structures

Based on what has been covered to date, a number of relationships can be suggested. These flow between the different types of contract and can be 'assessed' over a number of continua that indicate an appropriate structure for an organisation. The relationships between the vari-

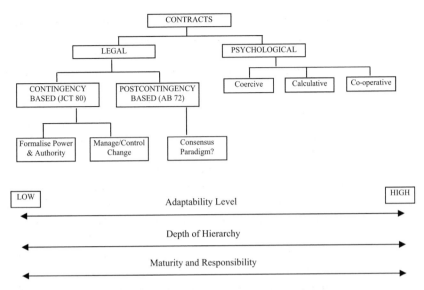

Fig. 5.1   Relationships between types and objectives of contracts.

ous types and purposes of contracts, and the suggested continua for structure assessment, are represented in Fig. 5.1. It is suggested that you spend a few moments considering what the figure is attempting to communicate before reading the following sections.

Individual readers will doubtless have interpreted the figure in slightly (possibly significantly) different ways, but there are a number of key 'messages' being transmitted and it is important to check that these are being received (to use the terminology of communication theory). The first message concerns the relationships between the types of contract: do not assume that relationships always fall clearly into the forms suggested here, as there will always be grey areas to consider. However, as a starting point, it is possible to suggest relationships between the legal and psychological types of contract. Coercive contracts, for example, can be strongly linked to contingency-based contracts on the basis of their common concern with power, authority and punishment/reward. Similarly, co-operative contracts can be linked with postcontingency contracts on the basis of their common concern with achieving/developing a consensus paradigm. This leaves calculative psychological contracts as not having a clearly identifiable link with any legal contract. Calculative contracts are suggested as fitting with both contingency and postcontingency contracts in that the less restrictive contingency contracts and the more restrictive postcontingency contracts will contain features of calculative psychological contracts. This then indicates that largely different structures will apply for each of the contract types, which may seem to cause a few problems with regard to the previous suggestion that individuals will tend to operate a range of psychological contracts at the same time.

Project managers have to be realistic – they cannot expect to meet all of the needs of each project team member by trying to run a range of psychological contracts when operating on the contingency (transactional) approach. It would seem to be more possible to achieve this objective when operating on a postcontingency (transformational) approach, but even so the project manager needs to ask whether they truly need to achieve this particular objective. Be aware that the needs an individual seeks to meet through establishing psychological contracts do not all require to be met within the work environment. The social environment can also contain psychological contracts, unless an individual is working so many hours that they do not have a social life!

The relationships between contract objectives and the assessment continua strengthen the argument for differing structures for differing projects. However, the continua used in this section require some initial discussion. The issue of organisation adaptability is suggested here as a valid indicator of how willing and able an organisation is to refocus itself on new objectives. The industrial-era paradigm is generally accepted as seeking high levels of control and consequently being unable to adapt rapidly, if at all, as the emphasis tends to become one of attempting to make the wider environment adapt to the organisation. Level of adaptability is therefore an important indicator of the required type of structure. Depth of hierarchy further identifies organisation structure, but in an explicit manner: is the hierarchy deep and multi-layered or is it shallow with few layers? An organisation chart will quickly illustrate hierarchy depth and it is generally accepted that rigid, contingency-oriented organisations, with their emphasis on issues such as increasing functional specialisation, will have greater hierarchy depth than an organisation which is postcontingency oriented.

Maturity and responsibility levels are probably the most acute indicators of whether an organisation is essentially bureaucratic or organic in nature. There is also some relevance in these factors, particularly responsibility at the personal level, for matrix and organic structures. Banner and Gagne (1995) suggest that the success of any such organization is dependent upon achieving a critical mass of human resources (people) who are in agreement with the adoption and operation of them. A significant problem in achieving this level of agreement is the traditional, Industrial Revolution paradigm with its emphasis on rigid hierarchies and increasing functional special-ism. Flexible and adaptable (mature and responsible) people do not generally result from this paradigm. Personal maturity, such as not taking the 'victim' approach (as in an individual not doing something because it is not within his or her job description), is argued to be es-sential if postcontingency organisation structures are to succeed.

The project manager therefore has to be realistic when developing the organisation structure for any new project, particularly if they have no input into the selection of contract form, as is frequently the case. An interesting exercise at this point would be to re-examine the issue of virtual teams covered in Chapter 4 and consider how the issues of legal and psychological contracts could interplay in such situations. Would you, as a project manager in such a situation, feel more comfortable with emphasising the legal or the psychological contract?

5.3.4 The law and organisation structures

There is no intention within this section to take a Big-Brother-is-watching-you type of approach – it is not intended to suggest that the implications of various laws for project organisation structures are either good or bad. On this basis, a statement that the general tone of legal implications is one of constraining the activity of individuals (how often do you see signs saying 'You are allowed to walk on the grass' as opposed to 'Please do not walk on the grass'?) should not be taken as inferring any judgement on the 'correctness' of the law – whether an individual chooses to ignore or obey a particular law is entirely a matter of personal maturity. However, assuming that laws will not be ignored then requires some consideration of the possible effects of those laws on the organisation structure. Basically, will the law constrain the structural possibilities by default?

A key starting point for this discussion is that, so far as the author is aware, there is no requirement within UK law, or any applicable European law, for organisations to adopt any specific *structure*. This statement requires some clarification, in that it is important to sepa-rate legal roles and responsibilities from the structure that an organi-sation chooses to adopt. For example, the issue of health and safety is increasingly seen as important, particularly perhaps with regard to the UK construction industry. Under current UK law and other Euro-pean Union (EU) legislation that has to be considered under the Single European Act, there is a wide range of responsibilities with regard to health and safety (H&S) which applies to all industries.

A particular construction industry example relates to the assessment of competence and resources with regard to H&S. The client for a particular project has an absolute duty to ensure that the parties responsible for design and production activities are competent to carry out a range of relevant activities (ECI 1995). Within this statement a number of roles (or functional specialisations) are implied, such as designers, planning supervisors, contractors, etc. Along with this is an explicit responsibility on the client to ensure that practitioners of such roles are competent. However, there is no legal requirement that the client has to carry out all the roles in-house, so it is entirely possible for the client, or its project manager, to organise the project in a variety of ways. Whichever way the project is organised should not, however, prevent the client (or its representative) from completing its responsibility for the assessment of competence.

In fact, two possibilities for organisation are offered by the legislation:

- Some or all of the roles are carried out within the client's organisation, in that the legislation refers to circumstances where roles are carried out by the client's employees and the resultant requirement to ensure that identified responsibilities are clearly assigned. A further requirement for the organisation structure to respond to is that staff have sufficient time and, where necessary, resources to carry out their duties. This possibility places requirements on the client with regard to aiding the achievement of competence (through the mechanism for assessment), but still does not explicitly state any requirements for organisation structure.
- Some or all of the roles are carried out by external organisations (consultants, contractors, agency staff, etc.). In such cases reasonable inquiries must be carried out to determine competence. Again, there is a mechanism for assessment identified, but no explicit requirements for organisation structure.

Even when the legislation specifically refers to the issue of organisation structure it is within the context of the client making certain that relevant rules for the achievement of safety, health and environment (SHE) requirements are embedded within the structure rather than any attempt to explicitly impose a structure. In section 8.3.2 of the ECI guidance document (addressing organisations and rules), there are listed a number of items. These are to act as the basis of a client's checklist concerning the review of organisational structure. Included items cover differentiating between the company organisation structure that is focused on general SHE (including those director(s) responsible for SHE) and the structure focused on project SHE. There also needs to be clear identification of line responsibility for SHE within both structures.

This chapter identifies legal requirements with regard to roles and responsibilities but makes no explicit statements with regard to organisation structure. This is not to say that the law does not imply structure – if a project manager could be shown to have organised a project in such a manner as to make the achievement of specific legal requirements difficult, or impossible, the legal system would doubt-less have something to say about it. However, so long as requirements are met, the general message seems to be one of leaving the structuring of an organisation to those directly involved. At this point we will discuss examples of how two organisations have dealt with the issue of structure in order to illustrate some of the possibilities, particularly with regard to creativity.

The first example considers a rather small project, the planning of a bid by the city of Anchorage, Alaska, to host the 1994 Winter Olympics. In many respects, the issue of organisation structure is relatively easily dealt with for a project such as this. The project is clearly defined, in that it has a finish point (cut-off date for applications to the selection committee), guidance is available for prospective applicants with regard to quantity and quality of resources/facilities required for a bid to be considered, and there are previous examples available to analyse for characteristics of successful and unsuccessful applications. Furthermore, the application is one subproject within the overall project (for the successful applicant) of actually running the Winter Olympics, so it need not dot all the 'i's and cross all the 't's with regard to that follow-on project. For example, the project team membership can be reconstituted if required for the follow-on project. It is arguable therefore that this particular project does not need to overly concern itself with achieving creativity and that the organisation structure can be highly contingency focused.

The structure of the bid's organising committee indicates that the project team did indeed focus on a contingency approach (see Fig. 5.2). Considerable effort seems to have been directed at identifying the roles and responsibilities that the team deduced would be required by the selection committee to be present in a successful bid. The organising committee's structure is very much a traditional (transactional) one in that it is composed of a deep hierarchy having many levels related to functional areas – there are seven functional areas at executive committee level and several of these have further subdivisions. Perhaps not the sort of structure that would be anticipated as being particularly flexible or adaptive, suggesting as it does

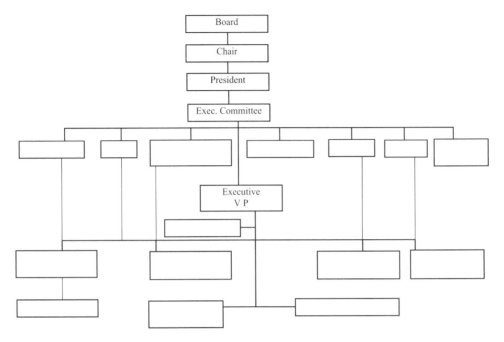

Fig. 5.2 Outline of organisation committee.

a tendency towards the use of rules and procedures. In fact, the committee was originally set up with the objective of engaging in a slow and deliberate project which would pick up momentum over time, having been set up in 1984 with the idea of placing a bid by 1989 to host the 1996 Winter Olympics.

Unfortunately, the project timeline was drastically reduced when the committee was asked in March 1985 to prepare a bid to be submitted in June the same year. This introduced a slight problem of having only 90 days for the whole process! It was at this point that the project changed in nature to become postcontingency oriented as the members of the committee, and the people of Anchorage, adopted the 'vision' of the project and began to ask what they could do to achieve its objectives. Various additional committees were formed to support the organising committee, a situation that could well have resulted in considerable conflict if these additional committees had become focused on mirage projects. Fortunately, the citizens of Anchorage seem to have exhibited high levels of maturity – there was significant public support, a large number of volunteers came forward, and the bid was completed in 30 days, with a further 45 being spent on developing the presentation. The bid was of a high enough standard to specific requirement added that clarity was required regarding who was to be responsible for the management of post-completion defects. Perhaps unsurprisingly, each of the four contractors came up with a different pro-curement route, with forms of contract ranging from traditional construction management through to design and build. This diversity of response was interpreted as evidence of a marketplace willing to take a wide variety of risk (some forms place more risk on the contractor than others), but that the product would be priced according to the risk accepted (Leitch 2001).

Bovis Lend Lease was ultimately awarded the project, and the form of contract it suggested was effectively an amalgam of several different ideas about the nature of the project and how that related to the company's recent experiences on other complex projects. The form of contract was probably closest to a traditional construction management procurement route, but had a number of significant differences. These were formalised through the use of a modified JCT 98 standard form and the progress of the project was such that there was a suggestion that the modified form should become a standard form in its own right, with a proposed name of PaddingtonCentral Building Contract. Praise indeed!

The modifications to the JCT 98 (without quantities) form include:

- open-book pricing of work packages;
- inclusion of a detailed work packages cost plan;
- contract sum based on cost plan plus an agreed contingency figure to represent the risk remaining in the project;
- a 'bonus' clause allowing Bovis Lend Lease to receive up to 50% of any savings on the contract sum;
- a 'penalty' clause requiring Bovis Lend Lease to contribute up to 50% of its agreed fee towards any cost overrun on the contract sum; and
- inclusion of a mechanism to adjust the contract sum in response to variations.

All of this seems to have met one of the client's initial desires: to create an environment in which everyone worked together to sort out problems arising during construction (which are inevitable) rather than being forced by the form of contract into defensive positions from Day One. It seems to be just a matter of being willing to take the risk of adopting creativity!

5.4 Conclusions

The legal system identifies a range of roles and responsibilities for an organisation but largely leaves the selection of the structure by which that organisation achieves the responsibilities placed on it to those individuals who are willing (and able) to accept the task. This can be a responsibility that not every individual (project manager or otherwise) is willing to accept. Various factors can be suggested as contributing towards such a situation. The issue of teams versus groups, for example, is one that raises a need to be aware of how the work environment around an individual constrains his or her freedom to act. Groups practise self-censorship as a result of social pressures against expressing divergent views – nobody is willing to stand out from the crowd.

A further consideration is the development of communities of practice within the project environment and the values that such communities adopt as a means to get the job done. An individual may, on paper, have very little freedom to act and yet the community of which he or she is a member allows them much greater freedom to find 'unofficial' solutions to problems. The various issues around the assignment of freedom to act are therefore suggested as being of considerable importance in moving forward from transactional structures of organisation and towards an alternative that will prove more capable of dealing with change in both the project's internal and external environments.

PART III
STRUCTURE FUTURE?

6 THE GENOME APPROACH

Non nova sed nove – not new things, but in a new way.

Introduction

Previous chapters have shown that contingency approaches to project management are essentially attempts to control projects through the development of back-up plans (contingency situations). These back-ups are then ready to be implemented in the event of a problem occurring that prevents the original plan from being completed. Unfortunately, no approach to planning a project is capable of identifying all the possible problems that may occur within a project's lifetime, nor is it feasible to work on such a basis – the cost and time implications of such a basis to planning effectively preclude it ever being implemented.

Because contingency approaches cannot identify all possible problems, it then becomes impossible to attribute probabilities to each possible problem so as to identify the most probable ones. This in turn prevents the development of a contingency plan for each of the most probable events. Likewise, it is not possible to produce meaningful contingency plans for all those unknown possibilities that do actually manifest themselves during a project's lifetime. These are the type of event referred to by Dainty and Moore (2001) as unexpected change events (UCEs) – an event that was not planned for and which has the effect of changing the project plan, irrespective of the scale of that change.

There is also the need to consider all possible opportunities that may occur during a project's lifetime – not every deviation from the programme is in fact a problem. Unfortunately, the contingency approach tends to emphasise UCEs as being problems before the possibility of any of them actually being opportunities is examined. This arises because of the perspective that time is money and once time has been spent it cannot be recovered – a rapid response is the best response. In short, the contingency approach can be constraining and uncreative when attempts are made to optimise a modern-day project's organisation structure through identifying and responding to risk.

A more flexible approach, and one possibly presenting a greater chance of success in responding to all probabilities and therefore optimising a project, is to mimic the genome in species development. What is to follow may seem ludicrous to those who are deeply wedded to the contingency approach, and it has to be acknowledged that much of this chapter is untried and untested material in the context of projects. However, that is not to say that the material does not represent a valid basis for future development of the approach to organising for projects. After all, most new ideas are initially met with scepticism, but some of them actually prove to be rather good! So, having got the warning out of the way, it's time to dive into the details, starting with a brief discussion of genomes.

6.1 The simplified genome

In the evolutionary perspective on life, genomes can be regarded as the means by which good characteristics within a species are rewarded by carrying them forward with the development of that species. As the environment around a species changes, the determination of what are good characteristics also changes. One example of this can be found in certain songbirds resident in urban areas of the UK (as stated in the earlier warning: some of what is to follow may initially seem ludicrous). As these urban areas have become more congested with people and all their activities, particularly noise from road traffic, it has been suggested that only those birds whose song can be heard above the background noise will succeed in attracting a mate. These birds will then pass on the 'loud' gene and the sound level of birdsong will incrementally rise over subsequent generations. So now you know one possible reason why birdsong seems to wake you up more frequently these days.

The flip side of good characteristics being carried forward is that bad characteristics, in effect, tend to be killed off. In the songbird example this would equate to all those birds possessing the 'quiet' characteristic not mating and therefore not passing on the characteristic to future generations. However, before you start to worry about the development of a super-species with no bad characteristics, which

- The other existing members of the team must agree to that person's appointment.

In this way, teams grow organically and it is usual for new teams to form on the basis of one or more associates having a new idea and setting up a team/manufacturing cell to investigate/produce that idea (golly, a project!). None of the teams has elected or appointed leaders – leaders are left to emerge, with the consensus of their fellow associates in the team, from the circumstances occurring at any given time. Likewise, teams are not composed of functional specialists as in the Industrial Revolution paradigm – each team member is trained for all of the tasks that a particular team requires to be completed. No company plant contains more than 200 associates.

An organisation such as Gore really has little in the way of a structure that would be recognisable by a traditional, contingency-oriented project manager, who would almost certainly worry about the lack of functional specialists and clearly identified responsibilities and rules. The only recognisable aspects of the organisation for such a project manager would be those required by the legal system the job titles of president and secretary-treasurer. However, such structures are not for everyone. As previously mentioned, they rely upon mature and responsible individuals and not everyone is willing to accept a psychological contract of such a nature.

Case study 3

It has been stated that creativity should be regarded as an important part of the problem-solving process when dealing with the issue of uncertainty in procurement. One example development in west London illustrates this point, in that both client and contractor seem to have exhibited some creativity in the procurement of a £86m package within an overall regeneration project valued at £350m (Leitch 2001). The regeneration project was focused on the redevelopment of the area around Paddington railway station. The project would result in over 730 000m² of new development. Phase 1 of the project was the £86m package mentioned. This contained two office buildings with over 32 000m² between them, a leisure unit of almost 4000m², over 6000m² of retail units, 210 residential units, and a landscaped public area. All of this had to be completed by summer 2002 from a start date in late 2000.

The client for this project took the unusual step of asking the four pre-qualified contractors who were to tender for it to suggest the most appropriate form of contract to use. The project objectives, or success criteria, were largely the traditional ones of time, cost and quality, but there was also a be selected from five competing bids by the United States Olympic Committee and was put forward to the International Committee. However, unfortunately for Anchorage, even high levels of maturity are not always enough to succeed, as the International Committee selected the French city of Albertville to host the games (Meredith & Mantel 1995).

Nonetheless, this project is a good example of how a contingency-oriented project and organisation structure can be quite rapidly converted to a project operating in a manner closer to the postcontingency approach. This can be done if the people involved are sufficiently mature and responsible to work within the spirit of an organisation structure rather than seeking to rigidly impose it. Simply because a structure looks transactional does not mean it cannot be operated as transformational – it is just harder to achieve.

The second example considers W. Gore and Associates, one example of an organic company – the company actually refers to itself as being built on what it calls *un-management*. Unlike the Winter Olympics committee example, consideration of Gore does not automatically constitute examination of a project. The company undoubtedly carries out projects, but in this example the consideration is of how an organisation can structure itself to survive in the long term. Two relevant factors emerge from this:

- there may be benefit in Gore's approach for very long-term projects; and
- forecasting long-term environmental conditions is generally considered to be almost impossible, so perhaps there is benefit in organising so as to be able to respond rapidly in the short term.

One unfortunate aspect of Gore's application of un-management, at least as far as this discussion is concerned, is that the company has no formal organisation chart, so readers are offered the opportunity to produce their own based upon their interpretation of the following information about the company. As Gore is an American company, the legal requirements upon it differ slightly from those in the UK. However, the only official titles within the company are those required by law, in this case president and secretary-treasurer – everything else is covered by what are referred to as 'associates'. No new associate can be employed until two criteria are met:

- An existing associate must act as sponsor, a role which includes finding work for that person to do.

proves capable of sweeping all before it, it is worth pointing out that there seems to be quite a bit of what may be referred to as redundancy within the behaviour of most species. It is not really appropriate to get involved in the old nature vs. nurture argument at this point, other than to state that there is still some uncertainty about how much of behaviour is pre-programmed (nature) and how much is learned from the environment (nurture). It is therefore difficult, as yet, to determine whether all behaviours are in some way pre-programmed (as the fight-or-flight mechanism seems to be) at the genetic level in the same way that eye colour is. It does seem safe, however, to suggest that good characteristics tend to be generic in nature, such as being able to identify anything in the surrounding environment that represents a risk. In this manner they are reasonably easy to pass on between generations, in that the individuals who are most able to recognise risk are most likely to survive longer and therefore may be presented with more opportunities to pass on the risk-identifying characteristic.

Unfortunately, being good at identifying risk may not be sufficient on its own to ensure your survival. There is also the need to make the correct decisions concerning how to react to that risk, and this reaction may be regarded by project management traditionalists as selecting the correct contingency plan, such as choosing between fight and flight. This still leaves one problem: in the real world, not everybody makes the correct decisions all the time. Even experts are not expected to get every decision correct, and from time to time the wrong decision will be made. Fortunately, even this may not mean that the individual is out of the game, as other generic characteristics may come to their rescue. This is, in effect, a case of whatever contingency plan is decided upon and implemented being adapted in real-time (an action that project management traditionalists may find deeply worrying, in that it seems to suggest to them that there is a lack of control).

For example, the decision to run away may be implemented when it would have actually proved better to stand and fight. However, the situation may be retrieved through the runner having more highly developed characteristics such as visual perception and cognition than the chaser. The runner is therefore better able to take advantage of opportunities (remember that not every possibility that enters the real world is automatically a problem) that present themselves in the environment through which he is navigating than is the chaser. It is important to appreciate that neither the runner nor the chaser may have been aware that a particular environment contains any opportunities prior to navigating it and therefore it would be pointless for either of them to have previously developed contingency plans for it.

Much better to adopt a postcontingency approach once the pre-programmed action has been implemented. There are only two types of player in this context: the quick (to identify opportunities) and the dead (who failed to identify opportunities).

The suggestion here is that the genome approach may be relevant to projects in a variety of ways, but all of them are argued to depend on the structure of the project allowing them to function. This is perhaps most clearly the case with the so-called learning organisations. These seem to have at least a hint of a genome approach within their structure, in that a prime purpose is for them to pass on what have been learned to be good characteristics to future generations of projects. The nature of learning organisations in the context of the genome approach is discussed further in Chapter 7, but please do not be tempted to read that section before completing the preceding sections, including the rest of this chapter. Unfortunately, the book has to adopt a reasonably linear format whereas in the real world there are advantages to being non-linear (all will be revealed later).

6.1.1 Introduction to Case study 4

In order to provide an environment where the proposed genome approach can be evaluated with the minimum of prejudicial preconceptions, a two-stage strategy will be implemented. The first stage is to detail a type of project team that is relatively new within project management but was introduced in Chapter 4: the virtual team. Stage two will involve easing into the genome approach by discussing what a simplified genome may be expected to comprise. Stage one should have the benefit of distracting those readers with prejudicial preconceptions regarding the use of a genome approach to project management structures, and giving them the opportunity to consider a project environment to which it may prove particularly suitable. They may then approach stage two with a more open mindset.

Case study 4

Within this section a case study will be analysed from the perspective of achieving the successful operation of a virtual team. Key factors will be identified in the summary. The names of the individual organisations have been changed so as to protect the guilty!

This case study comprises three separate organisations: Production Analysts Inc. of West Lafayette, Louisiana; Hoch Arbeit Gmbh of Limburg,

Germany; and Working Solutions Pty of Brisbane, Australia. These organisations have been aware of each other for several years and each has a good international reputation in the field of design for manufacture and assembly (DFMA). However, none of them had worked together previously. They were then presented with the opportunity to work together on a global project to develop a 'super tram' for public transport systems in major cities around the world. Their input to the project was to provide DFMA analysis of the tram's components as the design developed. They were not to undertake any primary design of either the tram or the required infrastructure, therefore the establishment of the boundary to their project was reasonably straightforward. This belief appears to have been a significant factor in them being sufficiently confident to decide that the project represented an opportunity to implement virtual teamworking.

Note: at this point you will probably start to form opinions as to how the relationships between the three organisations may have been managed. It would be useful therefore if you were now to make a few notes of any initial thoughts on this matter. You can refer back to these later in order to determine whether any significant changes in thinking have occurred after working through the case study. What factors would you see as being important at this point?

Hint: six factors are seen as being particularly important for successful virtual teams.

An important question regarding any completed project is: was it successful? The short answer in this case is no. There were some areas where good results were achieved, but the overall picture was less than encouraging. Costs were higher than anticipated, the project duration proved to be overly optimistic, and the quality of output (DFMA advice) was not generally up to the level required by the tram design team. While the project was not a total disaster, the three organisations involved came out of it with a reluctance to become involved in similar projects until such a time as they had put in place measures to deal with the problems that they individually perceived regarding such a working environment. Such was the extent of negative feeling about the attempt at joint working that the organisations made no attempts to compare notes on perceived problems. This should have been recognised as a problem, if only in connection with each organisation learning from its first experience of working in a virtual team environment.

In order to more thoroughly evaluate the strengths and weaknesses of the individual organisations with regard to the formation and operation of a virtual team, each organisation has been assessed against a checklist of success factors for virtual teams. The details of the assessment are listed below, but it is important to note one important description of the virtual team dynamic: '... virtual teams fool the organisation into thinking that the team members work together in the same space and time with the same set of organisational norms' (George 1996). In other words, as far as parent organisations are concerned, the virtual team is structured to produce an illusion. This is a situation somewhat like the earlier discussion of high and low entropy systems. Any failures with regard to the key factors will constrain the team's attempts to produce the required illusion and therefore mean that a virtual team has not been achieved.

Organisations can underestimate the need to plan and design around those differences that inevitably occur in virtual teams. These three particular organisations, in effect, fell at the very first hurdle as they failed to appreciate that issues involved in organisational design include the overall direction for the team, the structure it will adopt and the support systems available to it. Factors to particularly consider are:

- *Make efforts to define business goals for the team to work within.* Involving potential team members in developing business goals usually results in them concentrating on the needs of the whole project organisation rather than on pushing individual agendas.
- *Guide teamwork through team values that are behaviouralised.* These values need to be established and the manner in which each individual team member expresses their values through behaviour before the project goes live. This can require considerable input of effort, a factor which may incline teams towards having the minimum level of membership possible. Values that support multicultural, multifunctional work should be developed to overcome any gaps in members' values (George 1996).
- *Involvement encouraged through infrastructure.* Organisations should seek to develop an infrastructure that allows and encourages, rather than restricts and discourages, the involvement in the project of members of the virtual team along with any other interested members of participating organisations.
- *Team configuration and boundaries should enhance productivity.* Do not rely on management to deliver a configuration and boundaries for the team – the people who are involved directly in the achievement of the team's business goals will generally develop more relevant configurations and boundaries than those who are not. Furthermore, the designers of a team will always have the greatest level of ownership for its success.

None of the organisations involved in this project had experience of working in virtual teams outside of their own environment and naively assumed that the other organisations would either already be operating in a similar manner to them (with regard to 'in-house' virtual teams, which each organisation had experience of) or would automatically recognise their seniority and fall in with their way of doing things. Neither of these actually happened, and the resultant 'team' found it impossible to fully recover from this initial failure, which can be regarded as an example of each organisation not realising that the world of the virtual team member is one having no walls (Cantu 2000). Consequently, it is vital that the members' jobs are designed to be as tangible as possible so as to compensate for this. Issues that the three organisations should have structured their virtual organisation to consider include:

- *Realistic job previews*. An attempt should be made to define how the team member will spend their time and the environment in which they will function. In this manner, the member is more aware of the problems and opportunities they will be presented with and can therefore cope better. Each organisation had carried this out for its previous in-house projects, but it seemed to be automatically assumed that such an approach would also be sufficient for a larger-scale project involving external organisations.
- *Job accountability*. On small projects there may be a tendency to believe that the communication level required between team members effectively means that everybody is involved in everything. This is not the case and should be strenuously avoided when involvement in larger projects is being considered. Job accountability identifies in advance those jobs that require all team members' input and those that do not. Lipnack and Stamps (1997) suggest that a responsibility matrix will be sufficient for this requirement.
- *Issues of compensation*. A frequent problem with the formation of temporary teams is that individuals may feel that they are out of touch with developments in their particular specialism, especially within the 'parent' organisation. Virtual teams need to be particularly conscientious in measuring members on their ability to collaborate with others and operate on the basis of minimal, or no, supervision, and to make maximum use of their levels of specialist knowledge. Only Working Solutions Pty had any awareness of this need and attempted to reward its members of the virtual team on the basis of their contribution to that team rather than to Working Solutions.

Good team design will develop some of the earlier points further and add the following:

- *Selection of members*. This involves an initial decision concerning which is the more important: the person or the purpose. Typically, project managers tend to develop a list of people they want for a project. Virtual teams should emphasise identification of purpose first and then move on to the selection of suitable people.

- *Identity creation.* It is important to remember that a virtual team has fewer groundings in the 'real' world and therefore tends to seek an identity for itself more assiduously than other types of team. Team names are one way to do this. One example is that of a team at Sun Microsystems which has the rather wordy formal title of 'Sun Services Live Call Transfer Team', but has adopted the rather snappy name of LCT for internal use (Lipnack & Stamps 1997). People outside the team are unlikely to be privy to the life (identity) behind the acronym.
- *State the purpose.* This is a stage on from the identification of business goals mentioned previously. This is vital for virtual teams and its value cannot be overstated (Cantu 2000). The team should be seeking to produce a clear answer to the question, 'Why are we doing this?' In the case of the three organisations, the purpose was imposed from above by each of their management teams and the emphasis consequently varied between them.
- *Make total connections.* Virtual teams need to have available to them the maximum opportunities for communication between members. Typically information such as office location (postal address), phone and fax numbers, e-mail addresses, web page addresses and server names are seen as comprising the minimum level of opportunity. The addition of personal phone numbers and e-mail addresses, along with meeting places (for in-house subgroups of the virtual team), is beneficial to making total connections between members. This can mean that organisations have to reappraise issues such as internal security.

The approach of each of the three organisations to this aspect of structure was simply to issue their team members with basic contact details of a team leader in the other organisations. They were then left on their own to deal with making initial contact and the development of working relationships. This resulted in a lot of time being wasted as team leaders struggled to become aware of the identity and location of virtual team members in the other organisations. The organisation structure did not emerge in a coherent manner and it is arguable that the structure that did emerge was simply a 'Band-aid' fix that allowed the team to function but not in an optimal manner. Not a satisfactory way to operate and a surprising omission by organisations that appeared, individually, to have a good grasp of current information technology. They simply failed to appreciate that different technologies may be required when co-ordinating work through their respective external environments, particularly with regard to engineering face-to-face meetings.

While it may seem a contrary statement, it is a truism that virtual teams are considered to benefit significantly from face-to-face meetings between team members. The extent of such meetings should be around 25% of the total interaction between team members. However, this is not always possible and may in fact be seen by management as contradicting their (inaccurate) perceptions of how virtual teams should function. In cases where face-to-face meetings are either not possible or are prohibited by senior managers, interactive forms of communication technology become especially vital for the co-ordination of work between team members. Technologies such as video-conferencing (either traditional and/or desktop), groupware software which allows groups of individuals to work on given documents at different times, newsgroups, bulletin boards and intranets are all useful in aiding communication.

In the case of the three organisations, team members did manage to hold face-to-face meetings on several occasions, but these represented less than 5% of the known interaction between members (it is common for organisations to develop informal and less visible communication routes to support/subvert the formal and visible routes). Linked intranets were attempted but differences in organisation culture meant that, while technically acceptable, the team members found they were discouraged from using the links due to what they felt were inappropriate ways of using them by other organisations. These factors contributed to the situation where the majority of interaction took place through groupware and e-mail. This issue of interaction with other stakeholders was a particular problem due to the different software packages used in the DFMA analysis of design data by each organisation. In trying to resolve the problems of analysis based on the three DFMA techniques of Boothroyd/Dewhirst, Hitachi and Lucas, virtual team members felt that they were increasingly out of touch with others in their parent organisation. Likewise, each organisation became aware that its members of the virtual team were increasingly becoming focused on personal (in terms of trying to succeed in making the output of the software packages compatible) rather than organisational objectives.

Rather late in the project, each organisation tried to establish a form of interaction with its team members that tied them back into organisational goals. While any additional interaction may help to some extent, in this particular project it should never have been regarded as being anything other than a damage limitation exercise. One of the unfortunate outcomes was that other members of each organisation, who were not virtual team members on this project, became prejudiced against being involved in such teams in the future.

A continuation of the organisational goals issue can also be found in the need to structure the virtual project organisation so as to maximise success for individuals trying to achieve what is referred to as organisational re-entry. Re-entry into the parent organisation, or movement to another virtual team, can be an extremely disorienting experience for members of virtual teams. The process needs to be handled very carefully so as to manage the potentially conflicting requirements of the team members and their organisation – team members are typically concerned with moving successfully to a new project and with receiving recognition for their efforts in their previous project, whereas organisations are concerned with maintaining their advantage over the competition and may try to rush the re-entry process. If this happens, members may well, to use an aerospace analogy, burn up on re-entry. Not a pleasant experience.

Finally, there is a need for organisations to design re-entry schemes so that the organisation maximises learning from the experiences of the virtual team member. This can then be carried forward to future virtual projects and also addresses the member's concerns about compensation and career development. In the case of the tram project, both Hoch Arbeit Gmbh and Working Solutions Pty provided re-entry schemes that addressed the majority, but not all, of these issues to some extent. Even so, the experiences fed back to these organisations were largely negative, thus reinforcing the reluctance of either to participate in such projects in the foreseeable future.

Overall, the approach of the three organisations to dealing with the demands of virtual teamworking left a lot to be desired. This was largely because none of them had spent sufficient time in preparing for the exercise, largely due to their assumption that any previous experience of satisfactorily completing complex projects using in-house virtual teams would readily transfer to the larger and more complex (in terms of number of interfaces) virtual project. There was no significant consideration of how the project's organisation structure should be developed so as to reflect the different characteristics of a virtual project team operating in an environment that is in effect external to the environments of each individual organisation contributing to the team membership (it may be worth re-reading the section in Chapter 2 on system maintenance and regulatory activities). Unfortunately, virtual projects are significantly different in nature from traditional projects and require project managers who are willing to invest in both preparation for, and management of, the experience. One way to achieve these requirements is through the consideration of what can be referred to as simple genes (on the basis that more complex ones will follow later).

6.1.2 Simple genes

Nature seems to work on the basis of the most elegant solution being the simplest one that meets all of the requirements for a given set of circumstances. In this way life has been able to colonise some apparently very inhospitable parts of the planet. The great diversity of life-forms that we find around us has evolved from very simple beginnings, and the intention in this chapter is to examine how simple a genome approach to designing (possibly more accurate to regard it as evolving, but that may be too radical at this point) project organisation structures could actually be.

Looking at the human genome as an example of the evolution of one of the most open systems that we know of on this planet, we are faced with a complex and highly differentiated system (so far, so transactionalist) that is based on a DNA composed of only four components: G, A, T and C. Following this example rigidly would lead to the question: what would be the nature of the four components to be used in forming project-structure DNA? In the previous chapters a wide range of possibilities has been introduced, even though they were not then being explicitly considered as potential DNA candidates. It may even be possible to identify a DNA based on only three components (why use four if three will work just as well?). Possible candidates for this DNA triumvirate are L, P and M: labour, plant and materials. These form the most commonly used perspective on projects (that of resource requirements), particularly with regard to the planning function. However, are they just too transactionalist in nature? They have, after all, been around for a very long time without 'naturally' pushing forward the development of project organisation structures to the transformational model. This may simply be a case of, as with organisms in general, them not having been stimulated to bring about any change. Chapter 2 illustrated that for many thousands of years the environment external to projects has largely changed at a slow rate, and this could be taken as evidence that L, P and M are not incapable of achieving structure evolution: they simply have not had any incentive to do so. That situation is changing, so there is now a need to seriously question whether L, P and M can provide a project-structure DNA capable of evolving under the new environmental circumstances.

The issue of a project's external environment is obviously an important one. Frequent reference has been made throughout the previous chapters to the need for project structures that are capable of responding to unexpected, and sudden, change in their external environment. Perhaps then it would be possible to use L, P and M in conjunction with E (external environment). If so, this would bring the model up to the same number of components as human DNA, but only on the basis that E is capable of acting in conjunction with L, P and M. Unfortunately, this is not the case as the external environment rarely sets out to co-operate with a project. In the natural world, the environment external to an organism may present threats or opportunities to it but does so in a manner that cannot be regarded as a deliberate focusing of resources on that organism. If the organism's DNA is sufficiently robust, it will withstand the threat or seize the opportunity. In either case, the external environment is not really interested. On this basis, it seems that a further component is required in order to integrate the interactions between L, P and M so as to maximise the probability of survival. A possible candidate would be I (information). Again, previous chapters have shown the importance of information to projects: it is the glue that binds them together and it is standard practice for project structures to focus on gathering information concerning production resources (consider the example of MRP in this context). However, how relevant would the resultant genes be to the evolution of a project organisation structure?

6.1.3 Initial analysis for gene relevance to projects

Labour, plant and materials are obviously relevant to the success of a project, but are they relevant to a project organisation structure? Of the three candidates, only one has any freedom of choice, and that is the labour resource. Plant and materials resources are essentially guided by the labour resource. For this statement to be regarded as valid, the definition of 'labour' needs to be examined. In this context the labour resource is not seen to comprise simply the production operatives. It also comprises the management team members and may include elements of artificial intelligence. The issue of AI (and artificial life) has been discussed previously and there are examples of AI being used in decision-making situations for live projects. Put simply, this issue is one that will increasingly present project managers with a dilemma: are AI resources to be classed as labour or plant (on the basis that they may be regarded as being no more than intelligent machines)? Personally, I opt for AI to be regarded as labour on the basis that projects will increasingly allow it freedom of choice – there seems little point in developing rapid decision-making systems if they are then constrained by the need to have their decisions given the okay by a human. After all, if such systems are trusted with managing national missile defence systems, it seems only fair that they be trusted to make decisions about how much steel to order.

The issue of freedom of choice raises an important problem for our DNA candidate genes: if a gene has no freedom of choice with regard to responding to changes in the environment (it operates as a closed system), what use is it to the survival of a project organisation structure? This suggests that it would be more appropriate to seek out genes that have at least the potential for responding to environment changes. Such an approach would rule out the use of P and M as project-structure (hereafter referred to as the project organism)

DNA candidates, leaving just L from the original batch. While such a situation does have the undoubted characteristic of being a simple one, it is unfortunately too simple to get the job done. Genes need to provide an organism with variety in order to better respond to external changes, particularly so in turbulent environments. In human DNA four components can provide a wide diversity of possibilities: the human genome is huge and as yet we understand very little of the properties of most of it. However, if we take the perspective that G, T, A and C are all just shades of a common biology (or chemistry), a similar perspective could be applied to the remaining project organism DNA candidate of L.

The APM body of knowledge (BoK) was discussed in Chapter 4 and a question was raised regarding the suitability of such constructs to projects that are increasingly faced with having to become transformationalist in nature. One possible area of suitability could be with regard to the manner in which the BoK provides shades of a common knowledge amongst project managers. Six types of knowledge are suggested within the BoK: strategic, control, technical, commercial, organisational and people (APM 2000), the suggestion being that a suitable level of expertise in each of these knowledge areas would provide a project manager with the basis for successfully managing any project. This is on the basis that the knowledge areas have been identified to be generic across all projects. Given that freedom of choice has been identified as an essential criterion for any DNA candidate, the APM BoK can be regarded as providing six areas where freedom of choice can be exercised: S, C, T, C, O and P. Unfortunately, this results in two 'C' candidates, and as this would be confusing when linking the genes, it is suggested that the second one (commercial) be renamed as business (B), on the basis that the first listed knowledge area within commercial is 'business case'. These six candidates are suggested as providing a sound basis for the development of a project organism genome and this suggestion will be evaluated in the following sections.

6.1.4 Linking the relevant genes

Identifying potential genes for the project organism is an important first step, but an equally important problem to be addressed is that of how these genes are to be encouraged to come together in a manner that will optimise a project's chances of survival. In previous chapters we have discussed the sorts of problems that can arise when individuals come together in a joint undertaking (such as the development of mirage projects) and the differentiation that results as the transactionalist paradigm adds further complexity to its organisation structures in order to deal with the changing external environment. The linking process for our six candidate genes therefore needs to take a form that will avoid such problems. A useful exercise at this point is the examination of some of the assumptions that characterise the transformational paradigm. This may then point to the form of a possible linking process.

Banner and Gagne (1995) identify five assumptions within the transformational paradigm:

- Everything is part of one seamless whole and is therefore both connected to and influenced by everything else.
- The whole organises the parts through an inherent design and control function present in life itself. Everything operates in accordance with this function (Banner and Gagne actually refer to principles of cause and effect within design and control, but this presents possible problems with regard to non-linear systems and so I have referred only to the design and control function).
- We are designed to be co-creators with life, as evidenced by the forms around us being produced in our consciousness before being made 'real'. Most of us enter into a collective interpretation of the world around us, which we then refer to as reality.
- Choosing to adopt the design and control function of the whole will result in integration; choosing to ignore it will result in disintegration.
- We have reached the limits of what can be achieved through self-preoccupation. The cost of any gains achieved through it have been considerable: environmental pollution, increasing numbers of conflicts around the world, and so on.

The above assumptions will doubtless cause some feelings of unease – they perhaps all seem a little too 'New Age' for the majority. However, our changing understanding of the laws of physics governing not only the world around us but the whole universe (and others that we cannot yet see: Hawking 2001) has shown that there is at least a grain of truth in these assumptions. As scientific knowledge increasingly moves away from the transactional model of Newtonian science and towards the transformational model of Einsteinian science, it seems probable that the grains of truth will grow in number. This is all well and good, but how do these assumptions help with the problem of linking our candidate genes? The overall message seems to be one of not seeking to rigidly differentiate parts of the whole but to acknowledge the extent to which our six genes may interact with each other. In order to achieve this, we need to remove barriers to interaction that have been built up by the transactional paradigm's need to add complexity. We need to move away from linear relationships and towards non-linear ones.

Brodnick (2000) has suggested that the viewing of an organisation is a matter of perception in that the individual can choose to see a variety of forms individually but cannot see all of the possible forms simultaneously. Individuals cannot therefore be expected to view both the transactional and transformational structures for a single organisation – the process invariably involves the need to change their way of thinking. This essentially means that they have to stop thinking in a linear manner and start thinking in a non-linear manner. Brodnick argues that linear thinking actually confounds the processes (simplifying management, control large organisations, etc.) that it is supposed to be good at achieving. This is particularly so when combined with high levels of organisational and environmental complexity, and four particular limitations are identified:

- lack of adaptability;
- organisation members negatively impacted upon;
- linear interventions tend to be ineffective; and
- overoptimistic assessment of measurement and predictive abilities.

(summarised from Brodnick 2000)

Perhaps one of the more interesting statements made by Brodnick is that stability (in essence the lack of adaptability) is the path to organisational death (aka entropy). The suggestion then is that the organisation needs to facilitate connections between its subsystems (and with its external environment) that are dynamic (not stable) in nature. What is being sought is a form of dynamic equilibrium whereby the organisation can absorb turbulence. This form of structure also loses its internal linear definition (no hierarchies), but gains patterns of process in a similar manner to a living organism's ability to freely process energy (remember the release of energy by the human resource discussed in earlier chapters – 2 and 3 in particular?) and transforms inputs to outputs. Our candidate DNA components must be free to understand their role within the project organism along with the roles of others and thereby create a collective manifestation of that organism. This results in a fairly radical suggestion: the structure is whatever the players decide it needs to be at any given time. Such a suggestion may seem to a transactional mindset to be a recipe for anarchy. However, it must be remembered that transformational structures rely on the inputs of mature players and without them there will indeed be anarchy. This was the initial perception of a furniture manufacturing company discussed by Daft (2001).

The Rowe Furniture Company of Virginia had been organised along transactional lines for around 40 years when the decision was made to move to a more efficient form of organisation. The new organisation was intended to maximise the knowledge and experience of the production line workers. In order to do this most of the supervisory roles were eliminated, differentiation was removed (in a manner similar to that in Toyota, but with greater integration) and the workers were asked to form horizontal clusters prior to designing the new production environment themselves. Each cluster (again, not to be confused with Handy's use of the term) was given the freedom to select its own members, design its section of the production process and the production schedules. The resulting chaos was too much for some workers and they left (possibly just too transactional a mindset). Within a few weeks the chaos began to diminish, and as it continued to diminish productivity and product quality improved over the old levels. Daft (2001) suggests that the key to the success of this form of organisation was the existence of open access to information for all team members. Information that the old management structure had deemed inappropriate for dissemination to the workforce had now become freely available, thereby allowing the workers to react more rapidly than when they had simply waited for management to instruct them.

One way to approach the idea of such a free-forming structure is to consider the ecological model of organisation. The ecological model states that new organisations are always forming within the population (environment) and that these new organisations represent changes or variation from previous ones. Some of these variations will find an environment in which they can survive, while the remainder will die (consider the number of start-ups that fail: typically 50% of UK start-ups fail within the first five years). Of the survivors, a small number will grow to become large organisations (the UK construction industry for example is made up of organisations of which around 90% employ fewer than 20 people) and become institutionalised in the environment (Daft 2001).

Within the ecological model there are held to be two types of organisation: generalists and specialists. Generalists have a wide niche and offer a range of services, while specialists have a narrow niche and offer few services. The width of a generalist's niche protects it to some extent from environmental changes, whereas a specialist is more at risk if its narrow environment changes significantly. However, they do have the benefit of being able to respond to change faster because they are usually smaller. This then suggests that each of our candidate DNA components could be regarded as specialists operating within a generalist project environment. While this retains the transactional emphasis on specialism to some extent, it also begins to move towards the transformational model in that it requires some consideration of the complex relationships possible between the six 'specialists' within the 'generalist' project and external environments.

Within such a half-way house arrangement, there is the possibility of regarding the linkages as developments of the functional chimneys used in transactional models. Figure 6.1 illustrates the possible development from transactional, or cognate, structures to what may be referred to as transforming (not yet fully transformed), or semi-cognitive, structures. The figure should not be taken as suggesting that the transforming structure is in its optimum form, as it is in fact transiting through a phase in the development to a fully transformational (transformed) structure. The presented structure will be developed further through the remainder of the book in order to illustrate how a transformed structure may look.

6.1.5 The start genome

Having identified possible project organism DNA candidates, the problem of how to structure the start genome for the project then arises. A brief overview of the structure of the biological version of a genome is relevant at this point. The genome for a species is constructed from chromosomes that come together to form what is referred to as the haploid set. Chromosomes are, in turn, constructed from genes and DNA is the means by which genetic information is carried within the chromosome. This is what determines the colour of an individual's eyes and so on within the context of the genome that is increasingly coming to be viewed as determining everything that an individual will be capable of doing during their life. Returning to the concept of the ecological model of organisation results in the suggestion that because variations on the theme of organisation are always appearing, the genome should be a representation of that variation. Developing this suggestion further provides the heuristic that the genome can be based on the generic set of project management knowledge areas (our candidate DNA), but that the genetic material carried by the chromosome will be specific to the variation that emerges. So, for example, if the emerging organisation is a generalist

185

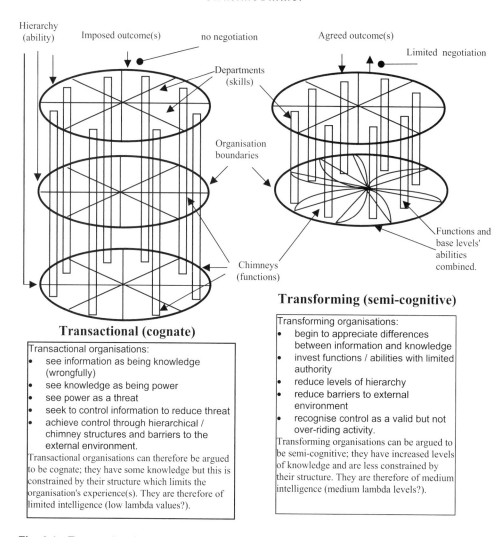

Fig. 6.1   Transactional vs. transforming structures.

one, the DNA information will be encoded so as to encourage abilities in generalist business activities. This may result in a greater emphasis on the control or business (commercial) areas than would be the case in a specialist organisation. Such a perspective is supported by the realisation that subcontracting organisations tend to be specialists, while main contractors tend to be generalists.

The ecology model therefore suggests that the proposed DNA of S, C, T, B, P and O could represent the basis of a viable organisation and that the project organism genome could be developed from it in one of two forms: the specialist and the generalist organisations. In the case of the specialist, the genome would perhaps be less loaded with regard to certain types of DNA than in the case of the generalist, and this loading would be represented by the linkages within the genome. As Brodnick (2000) has suggested, the transformational organisation can be seen as a complex nesting of dynamic systems and this nesting depends significantly upon the strength of communication within the whole. The genome should therefore allow for this communication in order to achieve the required level of system nesting. If each of our suggested DNA candidates is then regarded as a system, the genome illustrates how they are nested within the structure and is dependent upon information as measured by lambda. This point can be illustrated by considering how the start genome could be expanded.

6.2 Expanding the start genome

Taking the start genome to be S(trategic), C(ontrol), T(echnical), B(usiness), P(eople), and O(rganisation) presents the opportunity to expand it on the basis of the sub-areas of knowledge suggested for each of S, C, T, B, P and O within the APM BoK. However, simply taking the route of adding in the sub-areas as they are listed would add considerably to the DNA required to construct the project organism (it would add 39 'new' types of DNA). A more elegant approach would be to consider each of the sub-areas in the context of which of the DNA candidates they most closely represent. The sub-area of resource management, for example, could be argued to be most closely related to the technical area. Obviously, there is room for interpretation in this approach, and that is not seen as a bad feature. If nothing else, it will allow the opportunity for the genomes for generalist and specialist organisations to evolve into specific generalisms or specialisms. A suggested interpretation of the APM BoK categories is:

- Strategy – success criteria (T), planning (C), value management (B), risk management (B), quality management (T), health and safety (B).
- Control – work/scope management (O), time scheduling (T), resource management (T), budgeting management (S), change control (P), earned value management (B), information management (O).
- Technical – design management (C), requirements management (O), estimating (S), technology management (C), value engineering (B), modelling (B), configuration management (S).

- Business (commercial) – business case (S), marketing (O), financial management (C), procurement (S), legal awareness (P).
- People – communication (O), teamwork (O), leadership (O), conflict management (C), negotiation (B), personnel management (S).
- Organisation – life-cycle management (B), opportunity (S), design (T), implementation (C), hand-over (B), evaluation reviews (P), organisation structure (P), organisation rules (P).

(developed from APM 2000)

Such an approach gives an expanded genome of:

T,C,B,B,T,B/O,T,T,S,P,B,O/C,O,S,C,B,B,S/S,O,C,S,P/O,O,O,C,B,S/B,S,T,C,B,P,P,P.

An important point here is that there may be a tendency to view the suggested genome as being a linear entity; it is not. There is no intention to suggest any level of direct cause and effect within the structure of the genome. There are relationships within the chromosomes, but these are essentially non-linear, at least until further information is added. Also, there is no intention to suggest any aspect of the passing of time by the structure of the genome. The relationships illustrated will ultimately unfold in real-time, but as far as the genome itself is concerned, at this point they are simply relationships between and within chromosomes. To more fully illustrate this point, it would be quite acceptable to present the start genome in the following form:

T,C,B,B,T,B
O,T,T,S,P,B,O
C,O,S,C,B,B,S
S,O,C,S,P
O,O,O,C,B,S
B,S,T,C,B,P,P,P.

Using the expanded genome as a framework then provides an opportunity for individual organisations to develop a more specific project organism genome through the addition of further information about a particular project. Rather than plough through all six of the chromosomes within the expanded genome, a single chromosome will be taken for use as an example of how further information may be added to the genome. The example chromosome will be the organisation one, and this is selected on the basis of the previous assertion that mature individuals are important to the success of a transformational organisation structure; the organisation chromosome contains the largest quantity of 'people' DNA.

6.2.1 Adding people information

The three people DNA areas within the organisation chromosome concern the evaluation review, the organisation structure and the organisation rules. Dealing with the evaluation review first, this is suggested as being the sort of knowledge area where the competence of the individual is of considerable importance to the success of the activity. This is particularly important if the review is commenced as soon as the project goes 'live', as the primary intention here is to provide feedback to the project team. The issue of competence is one that has attracted more attention in recent years as the terms 'competence' and 'competency' have become increasingly fashionable to express the content of what should be the target of assessment and management development initiatives. A problem has arisen though because, while a consultant's credibility demanded the use of the word, few of them appeared certain what it meant. Also, the term competencies has come to be widely used but again in a confused manner. Partly this stems from the existence of yet another form of the term: competences.

While the dictionary definitions of these terms have retained stability of meaning, such as *The Concise Oxford Dictionary of Current English* defining 'competency' and 'competence' as 'ability (to do, for a task); sufficiency of means for living, easy circumstances; legal capacity, right to take cognisance (of court, magistrate, etc.)' and also regarding them as nouns with 'competency' and 'competence' being interchangeable, the social science meaning has changed from the dictionary definition. Current differences in meaning appear to result from the lack of a common consensus as to what these words represent in the social science context. In the managerial literature, however, subtle changes in emphasis can be found. For instance, competence may be defined simply as the ability and willingness to perform a task, and such a definition is broadly compatible with most usages of the term. However, for a past president of the American Management Association, competence was a more complex affair dealing with generic knowledge, motive, trait, social role or skill of a person linked to superior performance on the job.

Within the context of this section, competence will be regarded as an area in which an individual can be regarded as being competent (vital for the completion of a meaningful evaluation review). Competency will then be regarded as being knowledge and behaviour that supports competence; without competency an individual cannot be regarded as being competent. In other words, they need to possess the relevant skills, abilities, attitude and knowledge in order to be classed as competent. On this basis, the start project genome can be expanded by consideration of what competences, along with the required level of competency, are needed for a specific project. Again, these competences could be identified in generic categories based on the subject areas forming the project DNA, or a more bespoke and detailed identification could be used.

Organisation structure and rules are suggested as being very much people areas. Previous chapters have considered how the issue of trust within an organisation structure can have considerable effects on the attitudes of individuals and groups. The case of Toyota's early experiment with lean manufacturing is one example that we could usefully return to as a refresher. Another aspect of attitude that is relevant at this point has also been introduced previously: the issue of antagonism to the project by its players. D'Herbemont and Cesar (1998) suggested that individual players could significantly affect project outcomes through their adoption of antagonistic and synergistic attitudes to the project. This suggestion should be considered in the context of players expending energy in pursuit of project objectives: around 40–80% of project players are said to expend little energy in this way. Such individuals can be considered to be broadly antagonistic to the project, although it should be noted that many players exhibit both antagonistic and synergistic attitudes towards a project. Figures 6.2 and 6.3 illustrate how synergy and antagonism levels in project players may be measured. Such measurements may provide useful further information for the people DNA component of the project organism genome. These measurements would be particularly valuable if they were used at intervals during the project's duration to indicate the development of players in terms of moving from a transactional mindset towards a transformational one.

The issue of organisation rules is a particularly important one to treat in a non-linear, transformational manner. Organisations can become trapped in patterns of behaviour that are not conducive to long-term survival, and these patterns are enforced through rules, procedures, regulations, policies and so on. The dangers of teamthink have been discussed previously and should not need repeating here. However, an organisation needs the right kind of attitudes amongst its people to adopt ways of working that are not reliant upon reinforcement by rules and procedures.

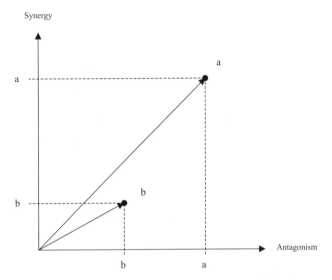

Fig. 6.2 Synergy and antagonism. (Source: Olivier d'Herbemont, Bruno Cesar, Tom Curtin, Pascal Etcheber, *Managing Sensitive Projects. A Lateral Approach*, 1998, Macmillan. Reproduced with permission of Palgrave.)

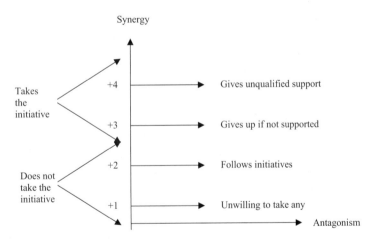

Fig. 6.3 Measurement of synergy. (Source: Olivier d'Herbemont, Bruno Cesar, Tom Curtin, Pascal Etcheber, *Managing Sensitive Projects. A Lateral Approach*, 1998, Macmillan. Reproduced with permission of Palgrave.)

6.2.2 Adding products information

We have mentioned previously that plant and materials information could not exhibit freedom of choice and were therefore removed from consideration as part of the project DNA. However, individual chromosomes may rely heavily on such (product) information. The control chromosome, for example, requires information on aspects of time scheduling and resource management, both of which could be significantly affected by the presence within a project of high levels of mechanised equipment, robotic or automated equipment, complex components and sub-assemblies, and so on. Projects of such a nature may then require their particular genome to reflect the input of techniques such as design for manufacture and assembly (DFMA). This type of technique requires the inputs of both specialist individuals and specialist products in the form of software packages such as Hitachi, Boothroyd/Dewhirst, etc.

The issue of product information also serves to raise the need to consider the connections between the components of the various chromosomes forming the project. When considering DFMA, for example, the structure will need to ensure that connections are made between the control, people, organisation and technical chromosomes along with any connections required to the external environment (remember: everything is part of the whole). This is one advantage of the genome approach in that it enables the identification of links that may otherwise be missed initially and then have to be added in a reactive manner when a related problem indicates that they are missing from the structure. Such a problem can occur when a parent organisation simply replicates itself in its external environment by imposing its own structure on the project. This could also be regarded as an example of parthenogenesis.

6.2.3 Adding parthenogenesis information

The dictionary definition of parthenogenesis is 'reproduction without fertilisation'; a type of reproduction common amongst primitive plants and so on. It can also be considered in the context of Maturana and Varela's (1980) suggestion that organisations create their own external environment. This was introduced in Chapter 4 and is worth returning to here as it raises an important issue for the project genome: if parthenogenesis is applied absolutely, there is no need for a project genome. Put simply, all project organisations would be structured in exactly the same manner as their parent organisations. Obviously, this is not the case, as shown by the ecological model. However, it can be a significant factor in that the ecological model also identifies the possibility of successful organisations becoming institutionalised within the environment (Daft 2001). At that point, they can delude themselves into thinking that they are immortal and seek to continue forward in time without changing in response to environment changes; they are the environment.

The issue of parthenogenesis is suggested as being of most significance when a project structure takes in new specialists (such as subcontractors) during the project's lifetime. Certain types of subcontractor in the UK construction industry have accrued a reputation for appearing to believe that the project revolves around them. This is generally due to them having a protected position within the form of contract – they provide a specialist service that is much in demand and they are few in number. The relevance of this *modus operandi* is that they can regard the project as an extension of their own organisation and have considerable negative impacts on areas such as conflict management, teamwork, technology management and so on. By considering the extent of parthenogenesis individual specialists bring with them, an opportunity to strengthen relevant chromosomes in advance is presented. One possible indicator of the level of parthenogenesis could be the use of antagonism measurement, although high levels of antagonism should not automatically be taken to indicate the presence of parthenogenesis. Further investigation of the environment may find it to be just one of possibly many factors in the presence of high levels of antagonism.

6.2.4 Linking the relevant genes

As information is added to the project genome, its length increases and the number of possible connections between the individual genes multiplies. The scale of the problems can be considered in light of the approximately 10 000 genes that are combined within the human genome. While it is not being suggested that even the most complex of projects would be expected to approach such a size, there may well be in the low thousands of genes in such a project. Each of these genes will need to be connected to at least one other gene. The basis of this suggestion is the interdependency issue that is part of the standard CPN approach to project planning: if any activity included in the analysis does not link with at least one other activity, its value to the project is zero and it should be removed from the analysis. Within the project organism genome there should be no genes that are not connected. Unlike the real world, there is no need to build in surplus capacity or redundancy to the project organism genome (POG) unless the project team can genuinely argue for the need to add entropy to the structure.

The nature of the inter-gene connections will vary from project to project. They will also vary during the lifetime of the project, both to respond to intended change and to deal with unexpected change. There is therefore no need to impose a transactional approach on the genome whereby all connections are determined in advance: connections should largely emerge as the project unfolds and this activity will be essentially the responsibility of the people DNA. An example of this can be found in Brodnick's (2000) assertion that leadership emerges in non-linear systems from the dynamic interactions between their subsystems. This involves the maximising of information flow through the system, and the POG should be capable of allowing this to happen. One possible form whereby the required connections can be encouraged to emerge can be found in the consideration of fractals.

Fractals are curious things in that they can exist at any scale within a system and yet they possess self-similarity throughout. This geometrical self-similarity is important in that if a pattern can be recognised at a very small scale, it can be taken as an indicator that the same pattern will emerge at some point in the future at a larger scale (McCauley 1995). In other words, large disasters have small beginnings. The trick is to recognise those small beginnings as soon as possible. There is, however, another suggestion for the use of fractals in the structuring of non-linear organisations. If the geometrical self-similarity rule is applied to teams, it is feasible that teams will seek to replicate their own structure within the environment of the project. The argument is that if a team can achieve self-management (that is, be successful), then so can the organisation of which the team is a part. However, in order for this to take place in a sustainable manner, there is a need to introduce feedback loops so that the self-management process achieves the required direction. These feedback loops should be of two types. Positive loops are important because they amplify behaviours that are providing beneficial outcomes; the project is on schedule, etc. Negative loops are important because they dampen behaviours that are providing detrimental outcomes; the project is behind schedule, etc. The linkages between relevant genes should therefore maximise information flow whilst incorporating both positive and negative feedback loops. At this point it would be worth revisiting the section on virtual teams in search of inspiration for the linkage process in the context of chromosomes.

6.2.5 The intermediate genome

As information concerning the project is added in the general manner described so far, the project genome moves away from its start conditions. However, it cannot yet be considered to have achieved its fully project-specific form and should be regarded as being of an intermediate nature. This process of development is of value in that it allows an opportunity to check the structure for emerging disasters. The previous consideration of fractals introduced the concept of geometrical self-similarity and this can be used as a checking mechanism at the intermediate stage of development. Self-similarity should allow for any errors introduced in the start genome to be identified, as they will have increased in scale by this point. The initial error may have introduced only a very small variation in the intended start conditions for the genome. However, even very small variations can produce disruption at a later stage. The intermediate genome also presents an opportunity to determine whether the feedback loops are operating as intended in that they should pick up any emerging detrimental behaviours in the project organism genome.

A key difference between transactional and transformational structures will be evidenced at the intermediate genome stage. Because of the reliance placed on people within the transformational approach, it will generally be found that the project team will require a wider constituency than players may have encountered previously. This increased membership may be regarded by some as being a measure of project effectiveness that indicates a detrimental effect on the efficiency of the project. After all, the more people there are involved, the greater will be the labour costs for the project. Such a perspective is in many respects an illusion. A proportion of the membership would have been involved in the development process anyway, and the remainder are being paid for their time irrespective of what project they are involved with. Such a negative perspective also fails to recognise the benefits of such early and wide-ranging involvement of players with regard to teambuilding.

It was previously noted that the sooner a project team can reach the stage in its life-cycle where it can be regarded as working synergistically, the better will be overall project performance. By involving a wider membership at an earlier stage, the genome approach begins the process, albeit somewhat loosely, of binding the ties between team members and thereby reduces the impact on team performance as individuals enter and leave the team during the project's lifetime. Remember that transformational organisations are essentially non-linear, so it is important to avoid the erecting of barriers to the involvement of any player who feels they have a contribution to make at any stage during the project.

6.3 Conclusions

This chapter has introduced some concepts and suggestions that may be difficult to accept. I could take the easy route and say that this difficulty is simply a reflection of individuals' transactional mindset. While this may well be valid to some extent, it would not be realistic to hold it responsible for all of the difficulty experienced. Much of what has been covered represents innovative material and in truth involves moving slightly beyond what would generally be regarded as a typical transformationalist mindset. While transformationalists refer to organic structures, these are not the same thing as structures that are organisms with a mappable genome based on project genes. It is only to be expected therefore that there will be problems in accepting much of what has been covered in this chapter. However, that is not to say that there is no benefit in questioning the material – seeking to understand non-linear concepts in general can be a useful exercise in personal growth as new ideas emerge from the provocation provided by new and challenging material. Speaking of which, the next chapter will provide more of the same!

7 TAKING UP THE OPTIONS

Medio tutissimus ibis – you will go safest in the middle.

Introduction

At some point the project has to move from the shelter of the project team's development process out into the real world with all its vagaries. A typical problem experienced by the transactional approach to developing a project has previously been noted as essentially being a desire to create the perfect project organisation structure. Unfortunately, almost as soon as this perfect structure meets the real world it is frequently found not to operate exactly as it was intended. The transformational approach takes a contrasting perspective in that perfection is not sought: an optimum structure is accepted as being the most that can be achieved. This optimum structure should not be viewed as being some sort of disappointing also-ran. Examples in previous chapters have shown that optimum structures can provide greater performance than their transactional predecessors.

Optimum structures must accept, however, that not all of the resources they would like to use are always available in the real world. Resources (plant, materials, labour) can be in short supply as problems are experienced elsewhere in the increasingly global market. A further problem is that such shortages will not always be constant over the duration of a project. There may be a shortage of a given resource at several points during a long-term project. At such times the project will need to fall back on its greatest resource: the people (in terms of their knowledge, expertise and motivation) within the project environment. Consideration of how a learning organisation functions can be informative with regard to the mistakes that can be made when an organisation forgets the knowledge that it (in the form of its members) has been exposed to in the past. In the event that things take a serious turn for the worse, part of the project member-ship's expertise and knowledge may be a few good jokes that can be used in an attempt to raise morale and spark a little more of that all-important creativity (so long as the project is structured in a manner that will encourage all of the above).

7.1 Required structure characteristics: discussion of Skunkworks

In order to further determine the required characteristics of a project genome, the operation of Lockheed Aircraft's Skunkworks team is worth examination. Skunkworks is the in-house name given to the members of the Lockheed Advanced Development Company (LADC), which was set up in the 1940s by Kelly Johnson. Johnson was aware of the increasing diversity of problems related to the de-sign of aircraft. Particular problems arose from the transactional ap-proach which had resulted in an assembly line approach to the design process; designers became more specialised, so the overall number of them rose and their knowledge of each specialism concerning the skills of other specialisms decreased. Johnson built his team with the intention of overcoming these problems and emphasised integration rather than differentiation. The result has been a successful design operation that has completed a number of important design projects over the years it has been functioning.

Some of the characteristics of the methods employed by Skunk-works players may be of relevance to the project genome development and so will be discussed briefly. Anyone wishing to read up further on this area could usefully visit MIT's website and search for reports on the Lean Aircraft Initiative, but be warned that the information does seem to be highly mobile.

The overall principle within Skunkworks is that of simplicity (LADC 1992). Every activity has to operate in the simplest possible manner, even in the face of pressure from the external environment, particularly the US government's apparent need for frequent meetings and reports, and increasingly complex forms of contract (remember the JCT/AB discussion in Chapter 5?). Skunkworks also recognises that it needs to achieve breakthroughs in the use of technology (much of it classified as secret), operate on the basis of low volumes and rates of production, but also get new capabilities operational quickly. As part of achieving these needs, Skunkworks has simplified its or-ganisation rule structure to what its members feel is the minimum to get the job done. Some of these rules do, it has to be noted, have a distinctly transactional feel to them and are therefore suggested as being worth evaluating with regard to a project genome intended to operate in a transformational manner. The number one rule, for example, is that the programme manager has complete control. A further rule is that the number of people connected to a programme in any way is rigorously restricted. This last rule can be contrasted with a further rule that Skunkworks and its customers must develop a relationship based on mutual trust. However, one rule that does have a transformational tone to it is the one stating that personnel must be rewarded on the quality of their work, irrespective of the extent of their supervisory responsibilities.

Ongoing developments at Skunkworks are intended to provide the infrastructure to support small teams of around 25 people (perhaps a little on the large size, but it seems to work in the context of designing and producing secret aircraft). The emphasis therefore becomes one of people (wetware) being supported in their use of information by hardware and software; the issue of open access to information is suggested as being crucial to the project structure and therefore should be considered within the project-specific genome.

7.2 *Achieving the project-specific genome (PSG)*

The emphasis thus far has been on the issue of information flow through the project structure. Information is deemed to be the life-blood of a transformational organisation and its structure must not only allow for the movement of information, it should actively encourage it. However, such an approach does present the possibility that there will simply be too much information flowing around and that the players in the project will be swamped by it, thereby impairing their decision-making ability. This suggestion is simply a reflection of the fact that the citizens of the developed world live in the information age where vast quantities of information flow around the system. There is, though, a difference between the quantity of information and its quality, with it being arguable that much of the information available is of little importance beyond fuelling the interests of hobbyists and journalists.

Assessing the quantity of information available is a fairly straightforward process, as shown by statements along the lines of the average Sunday newspaper containing more information than the average person in the 17th century was exposed to in their lifetime. On that basis, if an average person from the 17th century was transported to the present and put to work on a large project, there is little doubt that they would be swamped. Their knowledge, experience and expertise would simply not give them any points of reference to deal with the quantity of information flowing around them. However, most project managers have not recently been transported from the 17th century and their knowledge, experience and expertise do prepare them for the quantity of information flowing around them. Some even have the ability to select the useful information from the useless.

The project-specific genome should therefore be aware of the information needs of the project and of its players. A later section of this chapter considers how the project infrastructure can deliver energy to the project and it is certainly possible to consider information in terms of the energy required to find, retrieve, process and forward it within the context of a project. The quantity of energy should then vary as the information demands of a project vary over its duration, and also as information requirements change between projects. The project-specific genome therefore needs the ability to recognise constraints on, and variations in, the energy level of a project.

7.2.1 Effects of project constraints

Project constraints can simply be regarded as contributions towards the energy profile for a particular project. If it is not possible to use a particular method or material, for example, alternatives must be identified and the optimum replacement selected. The replacement may well require energy in a different quantity and/or at a different rate from that of the original, but that is the nature of a constraint. This is not to say that constraints should simply be accepted – the human resource does have some freedom of choice in such matters, hence the selection of an optimum replacement. It is also possible that the human resource may choose to expend some energy in being creative so as to identify a greater range of alternatives or to devise an innovative approach involving previously unused methods and/or materials.

The second way in which project constraints may affect the project's energy profile is through their unanticipated emergence in the project environment. In this type of scenario, the methods or materials that were intended to be used may prove to be unobtainable or, due to some change elsewhere in the project, no longer particularly suitable. Again, the human resource will need to be creative in finding alternatives to any resources that vary in quantity and/or quality from that intended. In order to provide the greatest opportunity for this to happen, the project structure should be capable of being expanded so as to deal with the problem by widening the project membership through recruiting new players on a temporary basis.

7.2.2 Life-cycle variations/identified genome phases

A significant benefit of a transformational organisation is that it can, in theory, reconstitute itself rapidly in order to respond to any sudden changes in its environment. Consequently, it would be expected to experience no significant difficulty in responding to changes that are a normal and expected part of its life-cycle. There are a number of models available to represent the life-cycle of both projects and organisations. Whilst all of these may appear to be different in terms of presentation, the number of factors considered, emphasis placed on individual factors and so on, they all have one common characteristic: they imply the need to solve problems in an acceptable manner.

The situation can become complex when the life-cycles of individual project players (contractors, subcontractors, consultants, etc.) come together within a project environment. The level of maturity of individual players would then be expected to vary and this may be manifested in their level of commitment to the project. The project genome should therefore be capable of responding to variability of maturity flowing from differing organisation life-cycles. However, care should be taken not to respond to this need by adopting a contingency-based approach which results in many different structures being planned in advance of natural life-cycle variations in the project's needs. Natural variations of this sort should not result in any UCEs and the transition between them should take place smoothly.

7.3 *Implementing the structure*

The project-specific genome should be a reflection of all the requirements with regard to energy release and the project's energy profile. In order to respond to natural project variations and UCEs, the structure must be fluid in nature. It is therefore suggested that the project environment should be regarded as being inhabited by many chromosomes of the S, C, T, B, P, O type. Each of these chromosomes will be required to link with at least one other chromosome so that a 'chain' may be formed. This chain will, in effect, be a representation of the project's required structure. In a simple project, the structure may be largely sequential, as shown in Fig. 7.1.

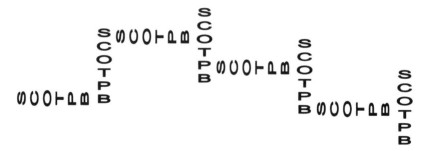

Fig. 7.1 Simple genome.

In a more complex project, the structure may itself appear to be more complex and less sequential. The comparison of a structure of this type with the type of bar-chart structures that would generally be regarded as being largely concurrent would seem to be a valid one. Structures such as that shown in Fig. 7.2 can therefore be referred to as highly concurrent genome structures.

The mechanism for establishing the linkages between chromosomes may appear rather simple on the basis of these figures. It is possible, however, that a specific mechanism for the guidance of link formation may be identified through a consideration of the fractal concept. Within the various characteristics of fractal physics there lies what are referred to as attractors. These are regarded as the means by which a system is bound to a pattern of behaviour such as

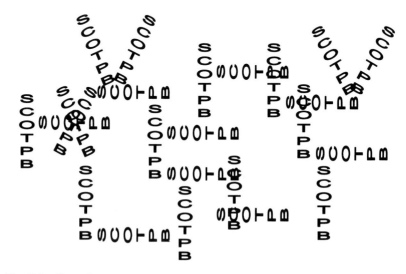

Fig. 7.2 Complex genome.

being drawn to a stable point in the environment or to an event having a regular cycle. A system may also be attracted to more complex forms of behaviour through the presence of what are referred to as strange attractors. These are attractors having more than one point of attraction within their finite space, and result in a system adopting behaviour that is classed as unstable but within certain bounds. Such behaviour is referred to as being bounded instability, as the system fluctuates in a complex but not completely unstable manner. It is, however, possible for a system exhibiting bounded instability to progress into completely unstable behaviour (Parker & Stacey 1994). These behaviours can be connected to the linkages illustrated in Figs 7.1 and 7.2.

Considering Fig. 7.1 first, it may be possible that the largely sequential nature of the chained chromosomes is a reflection of each having a single point of attraction to another. They can, in essence, be regarded individually as being attractors and the pattern formed is limited by each chromosome having only a single point of attraction. In the case of Fig. 7.2, a number of chromosomes exhibit multiple points of attraction, thereby presenting greater opportunities for links between them. The resulting pattern becomes less sequential and more concurrent. Such a pattern could be regarded as exhibiting the presence of a number of strange attractors amongst the chromosomes. The behaviour of such a system would be boundedly stable.

Taking this process a stage further, it is possible to envisage a project genome totally inhabited by strange attractor chromosomes, and the resultant pattern would be expected to be both highly complex and concurrent. The behaviour of such a system may become completely unstable, in which case the project system would 'die' (fail). It would also seem reasonable to suggest that in such highly concurrent structures there would be an equally high level of information required in order to establish the 'correct' linkages.

The consideration of information levels raises again the previously discussed concept of lambda as a measure of information, in which the optimum state is a lambda level of around 0.5. On this basis, it could be argued that the optimum form of project is one that supplies sufficient information to achieve a behaviour that is classable as bound stability. If the lambda level progresses towards the maximum level of 1, the system's behaviour may prove to become completely unstable. As with cellular automata, a project experiencing such a high lambda level would be expected to fail. Providing for the management of the flow of information therefore becomes a vital role for the project genome. This is illustrated in Fig. 7.3 which suggests how an

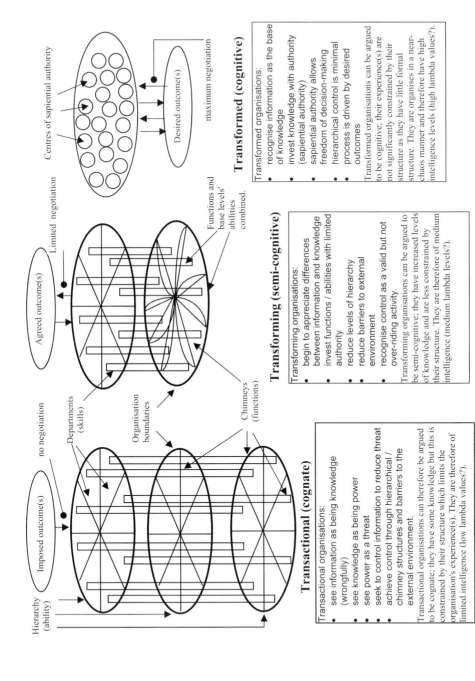

Fig. 7.3 Transactional, transforming and transformed structures.

organisation's structure may change as it moves from transactional through transforming to transformational.

7.3.1 Information flow management

One possible innovation is to consider the possibility of allowing for the use of emergent shapes. A basic representation of the concept of emergent shapes can be achieved through consideration of a simple bar chart. A novice may well initially regard the chart as a series of bars without being fully, or even partially, aware of what the bars may signify. Finke (1989) suggested that a drawing is in fact a structured entity composed of a visual image and an associated verbal (or textual) description of it. The bar chart can therefore be regarded as a structured entity. As the novice manipulates the visual image (activity bar length, location, relationships to other activities), new descriptions (structures) of it may emerge and it is these new descriptions that are regarded as an emergent shape. In essence, the chart may look the same in that the bars remain at the same length and in the same position, but the description of it changes with each new emergent shape. New descriptions may well include the concepts of specific interdependency between the bars (activities) and of discontinuity as a basis for separating the bars from each other.

The concept of the structured entity may prove to be of value in the process of managing information flows with regard to a project genome. An earlier section of this chapter considered the use of attractors and strange attractors as the basis for linkages between chromosomes. Such linkages could also be regarded as structured entities in that the pattern formed is both a visual image and a textual description. Working from the basis that much of the process of cognition in humans is based on searching for and finding patterns within the information presented to us almost continuously on a daily basis, it may well be that the management of information within the context of a project genome could be carried out through pattern recognition. By perceiving information in the context of being either complete or incomplete patterns, the human resource would be rapidly aware of where there were deficiencies in an environment formed of structured entities (and then be attracted to them?). Such an approach would allow all players within the project to achieve open access to all information throughout its duration. Some of the benefits that can be realised through the achievement of open access to information have been discussed in the context of virtual teams and the Rowe

Furniture Company's drastic improvements in productivity, quality and lead-in times.

7.3.2 Maintaining support

In order to maintain support for a project the systems activities of maintenance and regulation have obvious relevance, as discussed in Chapter 2. A further important consideration within this context is that of teambuilding. Considerable emphasis is placed within the transformational approach on the development of individuals in terms of issues such as their knowledge and competency, but of equal importance is the extent of their willingness to support the whole (the project) through their actions. A key aspect of this is their willingness to use initiative. This does, of course, depend upon the project structure giving them the freedom to exercise initiative. It is also important to recognise that individuals have a role in contributing to the team, and this aspect of maintaining support for the project can be a problem if a transactional approach is applied to team development.

Evidence of transactional thinking in respect of teams can be found in the emphasis placed on leadership of the team. Kehoe (1996), for example, claims that process improvement teams (used in the context of quality improvement) are critically affected by their leadership. In this context the leadership role is deemed to include actions such as selecting and inviting members, assigning tasks to individuals and holding members accountable. Such actions should be considered in light of the suggestion that team members should be self-managing – they will set their own assignments and hold themselves responsible. Kehoe does, however, suggest that the team leader should adopt a supportive, rather than dominating, approach to the activities of team members. Part of this supporting role may simply be providing the opportunity for teams to turn into communities.

The earlier discussion of Wenger's work on communities of practice raises the possibility that what an organisation should be seeking to achieve is not the development of teams (which are themselves more productive than groups) but the development of communities. This seems to be particularly so if the intention is to encourage the practice of self-management and self-organisation. Senge *et al.* (1999) claim that the involvement of individuals in the process of community (of practice) building gradually awakens the realisation in them that their common knowledge and/or experience is a good reason for them to get together. They then take on responsibility for organising themselves without reference to agendas or hierarchies that define their relationship. This process of organising again returns us to the discussion of energy within a project structure, but this time from a different perspective.

D'Herbemont and Cesar (1998) suggested that the release of energy was governed by the levels of synergy and antagonism within a project's players. Senge *et al.* (1999) take a slightly different perspective on the issue of energy by referring to an energy profile that defines the 'offering' (the project). The suggestion is that the energy profile will vary over the duration of the project as demand for resources rises and falls. In order for the needs of the energy profile to be met, a suitable infrastructure needs to be in place. This is composed of two items: culture and medium. Dealing with the medium first, this is classed as being the non-human element by which energy is released. An example would be the use of mechanisation in the project. The medium can generally be regarded as the most closed subsystem within the 'offering'. The culture component is regarded as being the human element involved in energy release and is deemed to be the most open system within the offering. The culture involved in achieving an offering (project) should be one of community building, with the emphasis being on the evolution of structure in response to perceived needs rather than it being imposed in some manner.

A good community should evolve a structure that will result in a successful delivery of the required energy profile. However, the project may fail for a number of reasons, such as inaccurate definition of the offering actually required and putting in place the wrong infrastructure. The infrastructure may be wrong because either the culture does not recognise that the offering has changed in nature or it fails to grow the optimum culture required. Part of that culture should be the ability to learn. Another part should be the ability to laugh while it is learning (see 'The humorous organisation' below).

7.4 The learning organisation

As with many other terms (see the previous discussion of competences and competency), the term 'learning organisation' is bandied around by all and sundry as the concept behind the term goes in and out of favour, while not always being understood. Various definitions can be found in the literature. Daft (2001), for example, refers to a learning organisation as being one that promotes communication and collaboration so that it may continuously experiment, improve and increase its capability through its members engaging in the identification and solving of problems. After the problems have been solved, it seems that some organisations suffer from a further problem to which they appear to be blind: what to do with the answers. This problem arises because of a simple fact: organisations are not living entities, therefore they do not have memories. The process that a learning organisation must engage with then becomes one of two stages.

In the first stage learning is achieved through the identification and solving of problems. Solutions that work and solutions that fail are both forms of learning, and the project will continue to provide opportunities for both until it is either completed or it 'dies' prematurely. The issue of creativity with a project genome has been discussed previously, so there is no need to dwell further on this in the context of problem identification and solving. The second part of the learning process has not been discussed previously and this relates to what may be referred to as 'organisational memory'. Pieters and Young (2000) note that an organisation's memory comes in two forms: first, in the recollections of its human resource and second, in whatever system of record-keeping/archiving that the organisation uses or has used in the past. The main problem that an organisation experiences when it relies on its human resource as its memory is that the resource is mobile. It can retire, get a job elsewhere or fall ill just when you need to access it. In such circumstances, the memory is lost. Similarly, when knowledge is stored in records such as files on a PC, the lost memory problem arises, but in such instances the memory is lost simply because nobody knows that it is there. This problem is sometimes referred to as Rembrandts in the attic, and this will be covered in a later section.

The project organisation therefore needs a structure driven by a genome that values organisational learning and memory. Various techniques are available to encourage the identification and retention of an organisation's knowledge. Knowledge mapping is one example. Pieters and Young (2000) provide a further example in the computer system developed by accountancy firm Price Waterhouse (as it was known at the time). This system allowed individuals with a particular problem to place a question about that problem onto bulletin boards accessible by every member of the organisation, irrespective of their geographic location. Almost overnight, it was then possible to tap into the experience and knowledge of thousands of players. It is not clear, however, whether the questions and their answers were then compiled into a database for later use.

Further possibilities are emerging as the medium of the project and organisation infrastructures become increasingly capable of delivering high-grade computing solutions to the problem of organisational learning. One example of such a solution is the emerging technology of ontological engineering with its claimed potential of being able to represent concepts (as opposed to knowledge as simple facts) generated within an organisation. It then becomes theoretically possible to produce a true knowledge map of an organisation that could be manipulated automatically rather than the current approach that is for such knowledge to be informal, in that it is located within the minds and documents of individual employees. This informal knowledge then must be interpreted by other employees who try to use it within the context of their perception of the organisation. A knowledge map would grow over time as changing knowledge was added to the unchanging knowledge about an organisation and it would ultimately be possible to reason with this knowledge, rather than simply collect it. Perhaps then the organisation would become a 'live' organism able to make its own decisions about how to develop in the future?

Until then, the project genome will have to content itself with the formal recognition of the learning problem by the inclusion within the project structure of a knowledge manager. However, as this is a newly emerging specialism (perhaps this will prove to be yet another example of complexity being added to the system?), there are no generally agreed formal requirements for the job, with experience being seen as the main qualification. Farr (2000) suggests that there is actually a need for two types of people to be involved in the management of knowledge. The first is a facilitator who can encourage the organisation to engage effectively with the learning process, and the second is a curator who manages the resulting output. That way, perhaps the Rembrandts will not be forgotten.

7.4.1 Rembrandts in the attic

Humans, along with many other living organisms, tend to have a few problems as they age. One of the more significant ones with regard to the subject of learning organisations is that we tend to start forgetting things, such as where we left the car keys and so on. Unfortunately for organisations there is evidence that they also experience this problem as they get older, and this can be referred to as the Rembrandts in the attic syndrome. Over time, organisations produce and process a considerable amount of information, some of which then becomes knowledge. Daft (2001) notes that as organisations become more successful and grow larger they also tend to become institutionalised. Part of this process would seem to be an increasing inability to tap into previous knowledge. Two examples of UK institutions which have suffered from 'Rembrandts' should illustrate the point.

BT (British Telecom) was, at the time of writing, involved in a legal battle. There would normally be nothing too unusual in that for a large organisation such as BT, but in this case it was, in a manner, seeking to derive benefit from a mistake that it had made. The mistake in question was that the company had forgotten something that it had done (it claims) in the past. While the legal niceties of the case may be of interest, they are not of importance in the context of this section. What is important is that the case illustrates the extent of missed opportunities that can arise when knowledge or information is forgotten.

In this case BT claims that it produced an invention several decades ago. The invention perhaps caused a brief flurry of interest at the time, but it was then rapidly forgotten by all involved. Fortunately, there was a record made of it and this surfaced a year or so ago (perhaps someone was doing a little housekeeping?) and shortly after that the full importance of the benefit was realised. BT is now arguing that without its invention, the internet (at least in its present form) would not have come into existence and that it deserves some recognition of this by extracting a payment from all those who are currently using the technology the company claims to have invented. The outcome of the case is potentially worth millions to BT, so perhaps it really did find the equivalent of a Rembrandt in its attic of a filing system.

The second example may not prove to involve as much money as in the case of BT, but it does illustrate that other benefits can flow from managing an organisation's knowledge effectively. In this example, the aerospace company BAe was trying to find a solution to a problem. The problem in question was how to formulate a paint that could be used on the external surfaces of aircraft. Nothing too radical there as there are many aircraft flying around covered in paint. However, BAe felt that its paint was special and would provide all sorts of benefits if it could just get it to work as intended. Much effort was expended on trying to get the formulation right before some bright spark had a wonderful idea: why not see if any of BAe's competitors had come up with a solution to the problem? The easiest way of doing this was suggested to be contacting the UK Patents Office and asking the people there to search their files for relevant patents (they will quite happily do searches in exchange for a fee). A few days later there was some good news and some bad news from the Patent Office. Yes, there was a company with a patent and it gave all the information that BAe needed to solve its problem. What is the name of that company, BAe asked. BAe, or at least an earlier incarnation of BAe several decades earlier, was the reply. Slightly embarrassing, but at least it could stop trying to solve the paint problem – again!

The Rembrandts problem is perhaps most significant for large organisations, particularly those which have grown through acquiring other organisations where it is difficult to establish fully the knowledge resource of those acquisitions. Nonetheless, the problem can occur at the project level when individuals who may just have the answer to a problem that has arisen are either not aware of the problem or are excluded by the structure from contributing their knowledge. Again, this raises the issue of achieving open access to information for all stakeholders in a project.

7.5 The humorous organisation

A final point to consider in the development of a project-specific genome is the relevance of humour. This may seem to be a rather strange subject to include, but it does seem to have an important role in the management process in general with regard to factors such as general level of motivation and in particular activities such as teambuilding. Anna Wardman, a former chief executive in the UK charity sector, once claimed that her management hero was an individual who had taught her to manage through the use of humour. A further example is the increasing amount of research being carried out on the role of humour in the context of management. At a conference on critical management studies held in Manchester in July 2001, papers were presented in streams with titles such as the future of work, constructing knowledge, and manufacturing futures. There was also an entire stream devoted to humour and irony, so the area does have some credibility!

A paper presented by Aufrecht (2001) gave an overview of many of the issues in the use of humour, amongst which he raised an interesting point with regard to the importance of humour in different environments. His initial observation was that a key function of humour is to allow the statement of things whose expression would normally not be allowed. Here is one example from the construction industry, but whose message is generic to any industry:

Q.   What's yellow and looks good on an architect?
A.   A JCB!

Having to explain that a JCB is a large piece of excavation plant sort of ruins the joke, so there is also an element of shared knowledge and experience within humour, and this can be of use in teambuilding exercises. There is, of course, the issue of culture within humour and the use of humour in multicultural projects needs a certain level of awareness of, and sensitivity to, cultural values in order to be successful. One example of this is that apparently there is no equivalent of the above joke in the Swedish construction industry. In fact, there is little evidence of any jokes concerning itself within the Swedish construction industry.

A further example of the relevance of humour can be found in the perceived lack of frivolity amongst the IO2 players in the collaboration on the MFD 01 project discussed in Chapter 2. The lack of frivolity was identified by IO1 players as a cultural difference that caused some problems. This does not mean that humour, either present or absent, is automatically a problem within projects, however. There are project managers who view a sense of humour as being one of the most effective management tools available to them. Of course, it is quite possible that those on the receiving end of the humour may not agree, and there do seem to be two clear camps with regard to whether humour is an appropriate management tool or not. Aufrecht (2001), for example argues that humour is most appropriate in organisations that are organised as closed systems. The more open the system, the less the organisation is suggested as needing to make use of humour. In the context of a transformational project organisation, the suggestion would be that there is no need to ensure a humour specialist is a player in the project, but there is a need for managers (or, more accurately, leaders) to be aware of the humour that will be present within the project environment and to identify the hidden messages that may be so important to the release of energy. The issue is therefore worthy of further investigation, even if it is only so that a few jokes can be worked in!

7.6 Conclusions

This chapter has covered some difficult areas of theory and has been able to provide only guidance as to how these areas may be applied in the real world rather than any rigid rules and instructions. Such a situation may cause some feelings of disappointment for anyone who was expecting to be handed a complete 'how-to-do-it' guide to developing a project structure genome. While this might be unfortunate, it is perhaps inevitable given that there are few fully transformational organisation structures around that can be used as test-beds. The process of moving from a transactional mindset to a transformational one does not happen overnight and it would be irresponsible of me to encourage individuals to rush out and restructure their organisations to become transformational on the basis of this chapter's content. Previous chapters have identified some of the problems that any form of structure will need to address, and their content also needs to be considered as part of the movement towards being transformational. In addition, the future holds further challenges, some of which are discussed in the final chapter.

8 FUTURE CHALLENGES

Felix qui potuit rerum cognoscere causas – fortunate is he who has been able to learn the causes of things.

Introduction

The emphasis in the preceding chapters has been on considering how the problem of developing relevant organisation structures for projects operating in an ever more rapidly changing environment could be dealt with. The majority of the work was focused on projects either in the past or ongoing in the here and now, with only a small element considering the longer timescale. This situation should by now be understandable in the context of the difficulty of predicting the future; a general heuristic is that the further into the future you wish to forecast, the less accurate the results will usually be. However, this chapter presents an opportunity to look at some of the problems that a project manager could face on projects in the future.

Given the problem of forecasting the future accurately, this chapter suggests six specific problem areas that could have a general (in some cases potentially significant) impact on the future role of a project manager. These problem areas are considered over three timescales: short (less than two years), medium (two to five years) and long term (five-plus years). At this point, it has to be acknowledged that such timescales are essentially arbitrary. Given that Moore's Law (a double of computing power every 18 months) is generally (and quite incredibly) acknowledged as still being valid, there could be significant breakthroughs that will shorten the timescales indicated. Of course, the whole timescale suggested could be completely wrong and even with many significant breakthroughs, some (or all) of the problem areas indicated may continue to be problems for many years to come. Such are the joys of navigating an uncertain environment.

The suggested problem areas are:

- *Short term* – recognising that any organisation structure is simply one 'space' in a near-infinite number of possible spaces; putting in place strategies to examine some of the near-possible spaces.
- *Medium term* – recognising that the knowledge worker (having sapiential authority) will become a factor beyond the control of the transactional organisation; adoption of a fully transformational organisation which can more readily tap into the range of possible spaces.
- *Long term* – achieving the maximum flexibility of organisation so as to respond almost immediately (through restructuring) to threats and opportunities within an organisation's environment; examination of the possibilities for truly learning organisations (those which do not need people to hold the knowledge but are sufficiently 'alive' to learn themselves).

These problems are best considered as following on from the work dealt with in previous chapters, which could, in this context, be regarded as having concentrated on:

- providing tools for developing an initial project organisation structure; and
- identifying the extent of diversity for ways in which to implement a more specific structure developed from the initial structure.

This 'following-on' process should be regarded as being within the context of two quite ambitious objectives from the previous work covered:

- To provide tools for the selection of the most relevant structure for each of the project's phases.
- To advise on achieving the project structure genome.

The relevance and validity of these objectives will perhaps have been (and may still be) difficult factors for some readers to determine. For example, the issue of what may represent a project structure genome possibly still eludes you at present. If so, don't worry about it – remember that none of this will happen tomorrow, and the vagaries of futurology are such that it may never happen at all. So look on the bright side: it seems that our planet inhabits a particularly high-risk part of our galaxy with regard to the probability of a terminal contact with a particularly large lump of rock, so the project structure genome may never actually become a problem that you have to deal with!

The emphasis here will be on considering issues such as how the objectives of a project organisation may change over time (those short, medium and long-term possibilities for the future of organisation introduced earlier). The fact that a genome can be defined as the genetic material for an organism is perhaps best regarded as a background issue until later in the chapter, when we will return to it in more detail.

8.1 The short term

Future objective 1: recognising that any organisation structure is simply one 'space' in a near-infinite number of possible spaces.

Perhaps the majority of present-day (parent) organisations do not even consider this possibility as existing when they come to the selection for their offspring of (project) organisation structure. This is because they take the tried-and-tested routes of either:

- imposing on their projects exactly the same structure that they (the parent organisation) have; or
- using the structure which they have always used on their projects, even if it is different from their own structure.

However, there is always the issue of change to consider. If a parent organisation seeks to ignore changes in its external environment through seeking to control its projects' internal environments by imposing structures on them, then the clock is ticking and its own end is nigh. Of course, it may be lucky and the external environment may well change back to something that the imposed project structure is able to deal with. After all, Railtrack (a privatised railway company in the UK) was effectively re-nationalised through its failure to perform, so it could be argued that the British rail system perhaps went through a period of change which brought it back to somewhere close to the starting conditions in its environment. However, this circular route was achieved only at considerable financial cost, so its effectiveness is well and truly open to criticism.

There may be those who find this suggestion politically tempting and/or attractive in its simplicity. However, it is far too simple. A project's external environment is potentially a highly complex (in terms of interactions between components within that environment) and fluid (in terms of the components entering and leaving the environment over time) place to operate. An important step towards dealing with this scenario is to recognise that project organisation structures may also need to be equally complex and fluid. As the old dictum goes: fight fire with fire. However, adoption of this approach also means there is a need to be able to envisage a near-infinite range of project organisational structures whilst also appreciating that differences between structures in such a range may be minute. Such minute differences may have little or no impact on the operational efficiency of 'different' structures.

Future objective 2: putting in place strategies to examine some of the near-possible spaces.

Having accepted that there may be a near-infinite range of possibilities for project organisation structures, the project manager is then faced with the problem of what to do with all those possibilities. There may well be such a thing as too much choice. Nonetheless, all good models of the decision-making process consider the act of implementation to be essential: make the decision and then implement it. In the situation of having to make decisions amongst a near-infinite range of possibilities, the requirement to implement could well seem like an overwhelming task. Combine this with the cost of decision making, insofar as it is frequently suggested that there is a break-even point between the cost of analysing possible 'solutions' and the cost savings or efficiencies presented by any of them, and the task seems to multiply its level of difficulty several times over. Fortunately, it seems that this need not be the case.

The key words here are suggested to be 'near-possible spaces'. These three words offer the possibility of saving a considerable amount of effort. While there is, at present, no clear mathematical relationship between the two concepts of 'near-infinite possibilities' and 'near-possible spaces', there does seem to be a fairly precise question which some project managers (and Bob the Builder, well almost) use: can we build it? The 'it' in question, in this instance, is the project organisation structure under consideration. However, the question needs a little more fleshing out, in that any constraints (and there always seem to be some) on building the structure need to be identified. These constraints are typically identified in terms of the availability of resources.

At present, the results achieved by asking such a question seem to depend heavily on the expertise in problem solving of those to whom the question is put. Even so, it seems to be one possible solution to the selection and implementation problems, and its value could well be enhanced in the short term by putting some effort into developing a strategy for identifying and then evaluating all the near-possible spaces that are out there. This issue of expertise with regard to problem-solving raises a possibility for longer-term development, in that an organisation needs to learn if it is not to be constrained within a few narrow areas of functioning. If the organisation chooses to try to control its external environment, and therefore remove the need to achieve successful change, it may not need to indulge in learning. However, as far as the majority of organisations are concerned, there is a need to respond to changes in their external environment that they cannot control. Hence the need to consider the process of learning, along with the development and use of expertise.

8.2 The medium term

Future objective 3: recognising that the knowledge worker (having sapiential authority) will become a factor beyond the control of the transactional organisation.

The issue of identifying near-possible spaces has been established as having implications for the development and use of expertise. This can perhaps be linked with an increasingly common management concept: the so-called learning organisation. Whilst the possible form and indeed the actuality of such an organisation has remained largely elusive for many organisations, there may well prove to be benefit in striving towards its achievement (as discussed in Chapter 7). A step towards this would seem to be the recognition by an organisation that its people are employed to carry out tasks that require knowledge in order to be completed. The response to such a suggestion may well be that a particular organisation has always recognised this situation. After all, one of the biggest developments in improved production came during the Industrial Revolution when employees became more and more specialised as new functions/trades/skills emerged. Unfortunately, an organisation's encouraging of employees to become increasingly specialised is not the same thing as recognising their knowledge.

The transactional organisation can be argued to actively seek to constrain an employee's gathering of knowledge – what possible use could it be for someone to develop expertise outside of the limited requirements for their highly specific role in the organisation? One answer seems to lie in the concept of automation. It is perhaps ironic that the increasing specialisation of human employees has contributed to the creation of such small and fragmented areas of expertise that they present a viable route for their automation. Indeed, there are those who argue that the employees resulting from the drive to specialisation are little more than automatons themselves. This is particularly so with regard to manufacturing industries such as the car industry. However, projects are not the same as car manufacturing – they involve constant change and require the interaction of the people involved (although involvement of people does not seem to have done Toyota much harm).

These factors introduce the need for knowledge. People who are capable of working effectively in such environments are not always easily found, as the UK construction industry has discovered at frequent intervals during its history. Unfortunately, scarcity of knowledge has not always resulted in the improved recognition of those who possess the required knowledge. Up until the Renaissance both engineers and constructors were generally regarded as being the lowest social order. One hierarchy established in Florence, for example, had four levels, and construction industry workers, along with constructors of machines, were definitely on the lowest level. One strange exception to this was if you were a constructor of machines used by theatres (for raising curtains, etc.), in which case you went up a level: entertainment was rated more highly than building useful structures such as houses.

The risk seems to be that if a society recognises the knowledge of a particular group as being valuable, that group gains in authority. Some master masons during the mediaeval period, for example, were able to buck the system and became as highly regarded as minor royalty, and there were instances where a mason had gained so much authority that he was able to take away control of a project from the original client.

The transactional organisation actively seeks to separate authority from knowledge by placing it in the context of position within the hierarchy. Unfortunately, deep hierarchies do not respond well to rapid change, particularly if the change is constant and accelerating, as it appears to be in many modern industries. In short, the transactional approach does not generally suit projects well and so it must be replaced. The problem then becomes one of time: culture does not change overnight and so organisations need to buy enough time to complete the change away from transactional and towards transformational, hence the need to consider this problem as a medium-term issue. In recognising that its workers should be invested with sapiential rather than positional authority, the organisation begins to change and presents the possibility of recognising near-possible spaces on the basis of the knowledge held by its workers. It then needs to have sufficient flexibility to move into such spaces when they present an advantage over the competition or simply the possibility to more efficiently utilise resources. After all, the concept of sustainability is just one example of the changing external environment faced by organisations – just how much knowledge do you have regarding sustainability?

Future objective 4: adoption of a fully transformational organisation that can more readily tap into the range of possible spaces.

Flexibility to respond to change is an important feature of a transformational organisation. Some may argue that the nearest form to being fully transformational an organisation can hope to achieve over the medium term is that of what is referred to as being an organic organisation. This concept has been suggested by those involved in its development as being the norm towards which future organisations will strive and that it is particularly relevant to project-based industries (such as IT, knowledge and service industries, which by association could be argued to include construction and possibly low-production-run engineering organisations). If this is so, what are its key features? These can be summarised as:

- highly decentralised, with large numbers of autonomous work groups (which should develop into true teams);
- highly flexible;
- highly adaptive;
- will form and disband as required to meet the needs of the parent organisation; and
- will not rely on rules to hold it together but will depend on vision (a compelling purpose 'owned' by the members of the organisation).

Perhaps a key consideration for organisations seeking to achieve an organic structure is that it is suggested as being composed of mature and responsible people capable of working in small groups on a face-to-face basis. Two possible problems seem to flow from this suggestion:

- Where are these mature and responsible people going to come from? Is it possible for such a structure to grow its own people or will they have to gain experience in a different organisation structure elsewhere until such point as they can be deemed sufficiently mature and responsible?
- The emphasis on face-to-face work would preclude such organisations seeking to utilise virtual teams. This situation conflicts with the assertion that the organic structure will most likely emerge within the information industries.

Both of these problems may well not be of significance in developing organic structures in the medium term, especially if they are to be regarded as a stepping stone towards a more fully transformed organisation structure which can be regarded as a long-term objective.

A further possibility is the type of structure sometimes referred to as 'agile' or 'lean', but the agile forms of manufacturing in particular do not seem to present the opportunities for recognition of sapiential authority that a truly transformational organisation should. While representing some interesting possibilities, they will not be discussed further here. Of more interest as the timescale moves into the long term is the concept of the so-called 'chaordic' organisation.

8.3 The long term

Future objective 5: achieving the maximum flexibility of organisation so as to respond almost immediately (through restructuring) to threats and opportunities within an organisation's environment.

We touched briefly on the flexibility of organisation when considering the knife-edge organisation in Chapter 2, but of greater relevance within this context is the concept of the chaordic organisation. While this type of organisation is included as a long-term consideration for this discussion, such an organisation is claimed to presently exist. The sole example would seem to be the Visa organisation. This organisation's structure was devised by the person who is claimed to have coined the term 'chaord': Dee Hock. A chaord is defined as being an organisation that operates with the maximum chaotic behaviour in combination with the minimum hierarchical order to ensure stability. As a matter of somewhat surprising record, Visa first opened for business in 1970, so we are not exactly talking cutting-edge in terms of up-to-the-minute research, but Visa is still operating, so perhaps there is merit in the chaord concept?

Mr Hock was one of the individuals who seem to follow the ideals of learning valued during the Renaissance, a so-called Renaissance man in that he read poetry, philosophy and science: a good mix of the qualitative and quantitative. Perhaps this mix was the inspiration for his idea to develop an organisation structure that it is claimed to be easily influenced by change, while also being highly resistant to failure. It has been claimed that Visa, as a single commercial organisation having a structure which is self-organising, is possibly unique: more than 30 years old and still on its own – it must be getting lonely by now. Irrespective of its claimed uniqueness, it does seem to have a number of features that are claimed to be essential for the success of any organisation seeking to decentralise. These can be identified as being:

- independent agents forming a network;
- checks and balances built into a federalist structure;
- ability to balance co-operation and competition;
- ability to respond positively to unplanned innovations;
- behaviour which can be regarded as being self-policing;
- providing most power to the end-user rather than to the organisation itself; and
- composed of a hierarchy which is claimed to be flexible and fractal.

(summarised from Senge *et al.* 1999)

Whilst all of these points are individually worth considerable examination, available space means that only one of them will be examined in any detail here. Given the previous discussion of the knife-edge organisation, the most relevant on the above list is suggested as being the point concerning the flexible, fractal hierarchy. This is suggested as enabling the creation of new levels of hierarchy, with each level having similarities of shape and format (in a manner similar to that of a fractal). Each level is insulated from the others in that it is a close replica of them, so if one level of the organisation fails (if a new division proves unprofitable, for example), the remaining levels would simply subsume its work and carry on. This form of structure is also claimed to have the benefit of easing the passage of information, due to there being consistent 'rules' about how decisions are to be made and information is to be processed. At first reading, the characteristics of such an organisation structure actually seem to be quite transactional in nature. However, the full subtleties of the structure will emerge through further research. Anyone wanting to read more on this organisation structure should start by looking at pages 391–396 of *The Dance of Change* by Peter Senge *et al.* (1999).

Future objective 6: examination of the possibilities for truly learning organisations (those which do not need people to hold the knowledge but are sufficiently 'alive' to learn themselves).

This is, in organisation structure terms, the Big One, the Holy Grail of organisation development. How will it be possible to create an organisation that does not use people as knowledge stores? Remember that when referring to knowledge we are not referring simply to facts and data: nice, hard system, quantitative stuff. Knowledge is somewhat more subtle than that. Data can be stored and added to with relative ease by the modern organisation. Cheap and powerful computing systems have ensured that, but have they also taken the first steps towards being able to create knowledge, develop expertise even? Could the chess-playing computer Big Blue be an example of this first stage? Did it actually learn as it played against Kasparov, or was it simply a case of huge computing power being able to calculate vast numbers of possible moves (with a rigidly defined set of rules to govern them) and select the one that would ensure victory? Perhaps you would care to consider this one further, but remember that the emphasis is on the organisation structure learning, not necessarily on carrying out or experiencing the work directly – humans could still be retained for this. After all, it would not do for them to just sit around enjoying themselves all day, now would it?

1 GLOSSARY OF PROJECT MANAGEMENT TERMS

Abstract resource   Imaginary resource introduced so that its availability and activity requirement gives an extra means of control – for example, two jobs not being worked upon simultaneously in order to obviate an accident hazard.

Acceptance   The formal process of accepting delivery of a product or a deliverable.

Acceptance criteria   Performance requirements and essential conditions that have to be achieved before project deliverables are accepted.

Acceptance test   Formal, predefined test conducted to determine the compliance of the deliverable items(s) with the acceptance criteria.

Accrued costs   Costs that are earmarked for the project and for which payment is due but has not been made.

Acquisition strategy   Determining the most appropriate means of procuring the component parts or services of a project.

Activity   Task, job, operation or process consuming time and possibly other resources. (The smallest self-contained unit of work used to define the logic of a project. In general, activities share the following characteristics: a definite duration, logic relationships to other activities in a project, use resources such as people, materials or facilities, and have an associated cost. They should be defined in terms of start and end dates and the person or organisation responsible for their completion.)

Activity definition   Identifies the specific activities that must be performed in order to produce project deliverables.

Activity duration   Specifies the length of time (hours, days, weeks, months) that it takes to complete an activity.

Activity file   A file containing all data related to the definition of activities on a particular project.

Activity ID   A unique code identifying each activity in a project.

Activity-on-arrow network   Arrow diagram, a network in which the arrows symbolise the activities.

Activity-on-node network   Precedence diagram, a network in which the nodes symbolise the activities.

Activity status.   The state of completion of an activity. A planned activity has not yet started. A started activity is in progress. A finished activity is complete.

Actual cost   Incurred costs that are charged to the project budget and for which payment has been made or accrued.

Actual cost of work performed (ACWP)   Cumulative cost of work accrued on the project in a specific period or up to a specific stage. Note: for some purposes cost may be measured in labour hours rather than money.

Actual dates   Dates are entered as the project progresses. These are the dates that activities really started and finished as opposed to planned or projected dates.

Actual direct costs   Those costs specifically identified with a contract or project. See also 'direct costs'.

Actual finish   Date on which an activity was completed.

Actual start   Date on which an activity was started.

Adjourning   The last stage of teambuilding where the team disbands.

Advanced material release   A document used by organisations to initiate the purchase of long-lead-time or time-critical materials prior to the final release of a design.

AND relationship   Logical relationship between two or more activities that converge on or diverge from an event. Note: the AND relationship indicates that every one of the activities has to be undertaken.

Approval   Term used when an individual accepts a deliverable as fit for purpose so that the project can continue.

Approval to proceed   Approval given to the project at initiation or prior to the beginning of the next stage.

Arrow   Directed connecting line between two nodes in a network.

Note 1: it symbolises an activity in 'activity-on-arrow'.

Note 2: it symbolises a dependency relationship in 'activity-on-node'.

Arrow diagram   See 'activity-on-arrow network'.

Arrow diagram method   One of two conventions used to represent an activity in a project. Also known as activity-on-arrow or i/j method.

As-late-as-possible (ALAP)   An activity for which the early start date is set as late as possible without delaying the early dates of any successor.

Associated revenue   That part of a project cost that is of a revenue nature and therefore charged as incurred to the profit and loss account. Note: associated revenue differs from the capital element of the project in that the capital element is taken as an asset to the balance sheet and depreciated over future accounting periods.

As-soon-as-possible (ASAP)   An activity for which the early start date is set to be as soon as possible. This is the default activity type in most project management systems.

Assumptions   Statements taken for granted or truth.

Audit Systematic retrospective examination of the whole, or part, of a project or function to measure conformance with predetermined standards. Note: audit is usually qualified, for example financial audit, quality audit, design audit, project audit, health and safety audit.

Authorisation The decision that triggers the allocation of funding needed to carry on the project.

Authorised un-priced work Any scope change for which authorisation to proceed has been given but for which the estimated costs are not yet settled.

Authorised work The effort which has been defined, plus that work for which authorisation has been given but for which defined contract costs have not been agreed upon.

Automatic decision event Decision event where the decision depends only on the outcome of the preceding activities and that can be programmed or made automatic.

Backward pass Procedure whereby the latest event times or the latest finish and start times for the activities of a network are calculated.

Balanced matrix An organisational matrix where functions and projects have the same priority.

Bar chart Chart on which activities and their durations are represented by lines drawn to a common time scale.

Note 1: a Gantt chart is a specific type of bar chart and should not be used as a synonym for bar chart.

Note 2: see also 'cascade chart'.

Baseline Reference levels against which the project is monitored and controlled.

Baseline cost The amount of money an activity was intended to cost when the schedule was baselined.

Baseline dates   Original planned start and finish dates for an activity. Used to compare with current planned dates to determine any delays. Also used to calculate budgeted costs of work scheduled for earned-valued analysis.

Baseline review   A customer review conducted to determine that a contractor is continuing to use the previously accepted performance system and is properly implementing a baseline on the contract or option under review.

Baseline schedule   A fixed project schedule. It is the standard by which project performance is measured. The current schedule is copied into the baseline schedule which remains frozen until it is reset. Resetting the baseline is done when the scope of the project has been changed significantly, for example after a negotiated change. At that point, the original or current baseline becomes invalid and should not be compared with the current schedule.

Benefits   The enhanced efficiency, economy and effectiveness of future business or other operations to be delivered by a project or programme.

Benefits management   Combined with project or programme management, benefits management is the process for planning, managing, delivering and measuring the project or programme benefits.

Benefits management plan   Specifies who is responsible for achieving the benefits set out in the benefit profiles and how achievement of the benefits is to be measured, managed and monitored.

Bid   A tender, quotation or any offer to enter into a contract.

Bid analysis   An analysis of bids or tenders.

Bottom-up cost estimating   The method of making estimates for every activity in the work breakdown structure and summarising them to provide a total project cost estimate.

Brainstorming   The unstructured generation of ideas by a group of people.

Branching logic   Conditional logic. Alternative paths in a probabilistic network.

Breakdown structure   A hierarchical structure by which project elements are broken down or decomposed. See also 'product breakdown structure (PBS)', 'organisational breakdown structure (OBS)' and 'work breakdown structure (WBS)'.

Budget   Quantification of resources needed to achieve a task by a set time, within which the task owners are prepared to work. Note: a budget consists of a financial and/or quantitative statement, prepared and approved prior to a defined period, for the purpose of attaining a given objective for that period. (The planned cost for an activity or project).

Budget at completion (BAC)   The sum total of the time-phased budgets.

Budgetary control   System of creating budgets, monitoring progress and taking appropriate action to achieve budgeted performance. Note: a budget should provide the information necessary to enable approval, authorisation and policy-making bodies to assess a project proposal and reach a rational decision.

Budget cost   The cost anticipated at the start of a project.

Budgeted cost of work performed (BCWP)   The planned cost of work completed to date. BCWP is also the 'earned value' of work completed to date.

Budgeted cost of work scheduled (BCWS)   The planned cost of work that should have been achieved according to the project baseline dates.

Budget element   Budget elements are the same as resources – the people, materials or other entities needed to do the work. Budget elements can be validated against a resource breakdown structure (RBS). They are typically assigned to a work package, but can also be defined at the cost account level.

Budget estimate   An approximate estimate prepared in the early stages of a project to establish financial viability or secure resources.

Budgeting   Time-phased financial requirements.

Budget unit   The base unit for the calculation. For example, the engineer budget element might have a budget unit of hours. Since budget units are user defined, they can be any appropriate unit of measure. For example, a budget unit might be hours, pounds sterling, linear metres or tons.

Burden   Overhead expenses distributed over appropriate direct labour and/or material base.

Business case   Information necessary to enable approval, authorisation and policy-making bodies to assess a project proposal and reach a reasoned decision.

Calendars   A project calendar lists time intervals in which activities or resources can or cannot be rescheduled. A project usually has one default calendar for the normal workweek (Monday to Friday, for example), but may have other calendars as well. Each calendar can be customised with its own holidays and extra work days. Resources and activities can be attached to any of the calendars that are defined.

Capital cost   The carrying cost in a balance sheet of acquiring an asset and bringing it to the condition where it is capable of performing its intended function over a future series of periods. See also 'revenue cost'.

Capital employed   Amount of investment in an organisation or project, normally the sum of fixed and current assets, less current liabilities at a particular date.

Cascade chart   Bar chart on which the vertical order of activities is such that each activity is dependent only on activities higher in the list.

Cash flow   Cash receipts and payments in a specified period.

Cash flow, net   Difference between cash received and payments made during a specific period.

Champion   An end-user representative, often seconded into a project team. Someone who acts as an advocate for a proposal or project.

Change control   Process that ensures potential changes to the deliverables of a project or the sequence of work in a project are recorded, evaluated, authorised and managed.

Change log   A record of all project changes, proposed, authorised or rejected.

Change management   The formal process through which changes to the project plan are approved and introduced.

Change request   A request needed to obtain formal approval for changes to the scope, design, methods, costs or planned aspects of a project. Change requests may arise through changes in the business or issues in the project. Change requests should be logged, assessed and agreed on before a change to the project can be made.

Child activity   Subordinate task belonging to a 'parent' task existing at a higher level in the work breakdown structure.

Client   The party to a contract who commissions the work and pays for it on completion.

Close out   The completion of work on a project.

Closure   The formal end point of a project, either because it has been completed or because it has been terminated early.

Code of accounts   Any numbering system, usually based on corporate code of accounts of the primary performing organisation, used to monitor project costs by category.

Commissioning   Advancement of an installation from the stage of static completion to full working order and achievement of the specified operational requirements.

Commitment   A binding financial obligation, typically in the form of a purchase order or contract.

Committed costs   Costs that are legally committed even if delivery has not taken place with invoices neither raised nor paid.

Communication   The transmission of information so that the recipient understands clearly what the sender intends.

Communications planning   Determining project stakeholders' communication and information needs.

Completion date   The date calculated by which the project could finish following careful estimating.

Compound risk   A risk made up of a number of inter-related risks.

Conception phase   The phase that triggers and captures new ideas or opportunities and identifies potential candidates for further development in the feasibility phase.

Concurrent engineering   The systematic approach to the simultaneous, integrated design of products and their related processes, such as manufacturing, testing and supporting.

Configuration   Functional and physical characteristics of a product as defined in technical documents and achieved in the product. Note: in a project this should contain all items that can be identified as being relevant to the project and that should be modified only after authorisation by the relevant manager (includes documentation).

Configuration audit   A check to ensure that all deliverable items on a project conform with one another and to the current specification. It ensures that relevant quality assurance procedures have been implemented and that there is consistency throughout project documentation.

Configuration control   A system through which changes may be made to configuration items.

Configuration identification   Identifies uniquely all items within the configuration.

Configuration item A part of a configuration that has a set function and is designated for configuration management. It identifies uniquely all items within the configuration.

Configuration management Technical and administrative agencies concerned with the creation, maintenance and controlled change of configuration throughout the life of the product. Note: see BS EN ISO 10007 for guidance on configuration management, including specialist terminology.

Configuration status accounting Records and reports the current status and history of all changes to the configuration. Provides a complete record of what has happened to the configuration to date.

Conflict management The ability to manage conflict creatively and effectively.

Constraints Applicable restrictions that will affect the scope of the project or the sequence of project activities.

Consumable resource A type of resource that remains available only until consumed (for example, a material).

Contingency Provision of a margin (for example, within the funds available for the project, as a float within the initial project plan or in over-specification of product characteristics) so that project achievement may be optimised against the project objectives in the face of risk impact allowing for the cost and opportunity cost of the margin. Or more simply: the planned allotment of time and cost or other resources for unforeseeable elements with a project.

Contingency plan Mitigation plan. Alternative course(s) of action devised to cope with project risks. See 'risk management plan'.

Contingency planning The development of a management plan that uses alternative strategies to minimise or negate the adverse effects of a risk, should it occur.

Contract A mutually binding agreement in which the contractor is obligated to provide services or products and the buyer is obligated to provide payment for them. Contracts fall into three main categories: fixed price, cost reimbursable or unit price, but may contain elements from each.

Contract budget base   The negotiated contract cost value plus the estimated value of authorised but un-priced work.

Contract close-out   Settlement of a contract.

Contract target cost   The negotiated costs for the original defined contract and all contractual changes that have been agreed and approved, but excluding the estimated cost of any authorised, un-priced changes. The contract target cost equals the value of the budget at completion plus management or contingency reserve.

Contract target price   The negotiated estimated costs plus profit or fee.

Contractor   A person, company or firm which holds a contract for carrying out the works and/or the supply of goods in connection with the project.

Control   The process of developing targets and plans; measuring actual performance, comparing it against planned performance and taking effective action to correct the situation.

Control charts   Display the results, over time, of a process. They are used to determine whether the process is in need of adjustment.

Co-ordinated matrix   An organisational structure where the project leader reports to the functional manager and does not have authority over team members from other departments.

Co-ordination   The act of ensuring that work carried out by different organisations and in different places fits together effectively. It involves technical matters, time, content and cost in order to achieve the project objectives effectively.

Corrective action   Changes made to bring future project performance back into line with the plan.

Cost account   Defines what work is to be performed, who will perform it and who is to pay for it. Cost accounts are the focal point for the integration of scope, cost and schedule. Another term for cost account is control account.

Cost account manager   A member of a functional organisation responsible for cost account performance and for the management of resources to accomplish such tasks.

Cost-benefit analysis   An analysis of the relationship between the costs of undertaking a task or project, initial and recurrent, and the benefits likely to arise from the changed situation, initially and recurrently. Note: the hard, tangible, readily measurable benefits may sometimes be accompanied by soft benefits which may be real but difficult to isolate, measure and value. (Allows comparison of the returns from alternative forms of investment.)

Cost breakdown structure   Hierarchical breakdown of a project into cost elements.

Cost budgeting   Allocating cost estimates to individual project components.

Cost centre   Location, person, activity or project in respect of which costs can be ascertained and related to cost units.

Cost code   Unique identity for a specified element of work. (Code assigned to activities that allow costs to be consolidated according to the elements of a code structure.)

Cost control point   The point within a programme at which costs are entered and controlled. Frequently, the cost control point for a project is either the cost account or the work package.

Cost control system   Any system of keeping costs within the bounds of budgets or standards based upon work actually performed.

Cost curve   A graph plotted against a horizontal time scale and cumulative costs vertical scale.

Cost element   A unit of costs to perform a task or to acquire an item. The cost estimated may be a single value or a range of values.

Cost estimating   The process of predicting the costs of a project.

Cost incurred   Costs identified through the use of the accrued method of accounting or costs actually paid. Costs include direct labour, direct materials and all allowable indirect costs.

Cost management   The effective financial control of the project through evaluating, estimating, budgeting, monitoring, analysing, forecasting and reporting the cost information.

Cost overrun   The amount by which a contractor exceeds or expects to exceed the estimated costs and/or the final limitations (the ceiling) of a contract.

Cost Performance Index (CPI)   A measure, expressed as a percentage or other ratio of actual cost to budget plan. (Ratio of work accomplished versus work cost incurred for a specified time period. The CPI is an efficiency rating for work accomplished for resources expended.)

Cost performance report   A regular cost report to reflect cost and schedule status information for management.

Cost plan   A budget which shows the amounts and expected dates of incurring costs on the project or on a contract.

Cost plus fixed fee contract   A type of contract where the buyer reimburses the seller for the seller's allowable costs plus a fixed fee.

Cost plus incentive fee contract   A type of contract where the buyer reimburses the seller for the seller's allowable costs and the seller earns a profit if defined criteria are met.

Cost reimbursement type contracts   A category of contracts based on payments to a contractor for allowable estimated costs, normally requiring only a 'best efforts' performance standard from the contractor. Risk for all growth over the estimated value rests with the project owner.

Cost/Schedule Planning and Control Specification (C/SPCS)
The United States Air Force initiative in the mid-1960s.

Cost-time resource sheet (CTR)   A document that describes each major element in the WBS, including a statement of work (SOW) describing the work content, resources required, the time frame of the work element and a cost estimate.

Costs to complete   Risks of all cost growth rest on the performing contractor.

Cost variance   The difference (positive or negative) between the actual expenditure and the planned/budgeted expenditure.

Credited resource   Resource that is created by an activity or event and can then be used by the project.

Critical activity   An activity is termed critical when it has zero or negative float.

Critical path   Sequence of activities through a project network from start to finish, the sum of whose durations determines the overall project duration. Note: there may be more than one such path. (The path through a series of activities, taking into account interdependencies, in which the late completion of activities will have an impact on the project end date or will delay a key milestone.)

Critical path analysis   Procedure for calculating the critical path and floats in a network.

Critical path method (CPM)   A technique used to predict project duration by analysing which sequence of activities has the least amount of scheduling flexibility. The critical path method is a modelling process that defines all the project's critical activities that must be completed on time. The start and finish dates of activities in the project are calculated in two passes. The first pass calculates early start and finish dates from the earliest start date forward. The second pass calculates the late start and finish activities from the latest finish date backwards. The difference between the pairs of start and finish dates for each task is the float or slack time for the task (see 'float'). Slack is the amount of time a task can be delayed without delaying the project completion date. By experimenting with different logical sequences and/or durations, the optimal project schedule can be determined.

Critical performance indicator   A critical factor against which aspects of project performance may be assessed.

Critical success factor   A factor considered to be most conducive to the achievement of a successful project.

Criticality index   Used in risk analysis, the criticality index represents the percentage of simulation trails that resulted in the activity being placed on the critical path.

Customer   Any person who defines needs or wants, justifies or pays for part of the entire project, or evaluates or uses the results. Could be the project promoter, client, owner or employer.

Cut-off date   The ending date of a reporting period.

Dangle   An activity or network which has either no predecessors or no successors. If neither, it is referred to as an isolated activity.

Decision event   State in the progress of a project when a decision is required before the start of any succeeding activity. Note: the decision determines which of a number of alternative paths is to be followed.

Delaying resource   In resource scheduling, inadequate availability of one or more resources may require that the completion of an activity be delayed beyond the date on which it could otherwise be completed. The delaying resource is the first resource of an activity that causes the activity to be delayed.

Delegation   The practice of getting others to perform work effectively which one chooses not to do oneself. The process by which authority and responsibility is distributed from project manager to subordinates.

Deliberate decision event   Decision event where the decision is made as a result of the outcomes of the preceding activities and possibly other information but it cannot be made automatically.

Deliverables   End products of a project or the measurable results of intermediate activities within the project organisation.

Delphi technique   A process where a consensus view is reached by consultation with experts. Often used as an estimating technique.

239

Dependency   Precedence relationship. Restriction that one activity has to precede, either in part or in another activity. (Dependencies are relationships between products or tasks. For example, one product may be made up of several other 'dependent' products or a task may not begin until a 'dependent' task is complete. See also 'logical relationship'.)

Dependency arrow   A link arrow used in an activity-on-node network to represent the inter-relationships of activities in a project.

Design authority   The person or organisation with overall design responsibility for the products of the project.

Design and development phase   The time period in which production process and facility and production processes are developed and designed.

Deterministic network   Network containing paths, all of which have to be followed and whose durations are fixed. Note: deterministic network is a term used to distinguish traditional networking from probabilistic networking

Direct costs   Costs that are specifically attributable to an activity or group of activities without apportionment. (Direct costs are best contrasted with indirect costs that cannot be identified to a specific project.)

Discounted cash flow (DCF)   Concept of relating future cash inflows and outflows over the life of a validity to comparison of projects with different durations and rates of cash flow.

Discrete milestone   A milestone that has a definite scheduled occurrence in time. Logical link that may require time but no other resource.

Dummy activity in activity-on-arrow network   An activity representing no actual work to be done but required for reasons of logic or nomenclature. Note: there are three uses for a dummy activity in 'activity-on-arrow network': logic, time delay and uniqueness.

Duration   The length of time needed to complete an activity.

Duration compression   Often resulting in an increase in cost, duration compression is the shortening of a project schedule without reducing the project scope.

Earliest feasible date   The earliest date on which the activity could be scheduled to stay based on the scheduled dates of all its predecessors, but in the absence of any resource constraints on the activity itself. This date is calculated by resource scheduling.

Early dates   Calculated in the forward pass of time analysis, early dates are the earliest dates on which an activity can start and finish.

Early finish time   Earliest possible time by which an activity can finish within the logical and imposed constraints of the network. (The early finish date is defined as the earliest calculated date on which an activity can end. It is based on the activity's early start time which depends on the finish of predecessor activities and the activity's duration.)

Early start time   Earliest possible time by which an activity can start within the logical and imposed constraints of the network.

Earned house   The time in standard hours credited as a result of the completion of a given task or a group of tasks.

Earned value   The value of the useful work done at any given point in a project. Note: the budget may be expressed in cost or labour hours.

Earned value analysis   Analysis of project progress where the actual money, hours (or other measure) budgeted and spent is compared to the value of the work achieved.

Earned value cost control   The quantification of the overall progress of a project in financial terms so as to provide a realistic yardstick against which to compare the actual cost to date.

Earned value management   Earned value analysis. Technique for assessing whether the earned value in relation to the amount of work completed is ahead, on or behind plan.

Effort   The number of labour units necessary to complete the work. Effort is usually expressed in staff-hours, staff-days or staff-weeks and should not be confused with duration.

Effort-driven activity   An activity whose duration is governed by resource usage and availability. The resource requiring the greatest time to complete the specified amount of work on the activity will determine its duration.

Effort remaining   The estimate of effort remaining to complete an activity.

Elapsed time   Elapsed time is the total number of calendar days (excluding non-work days such as weekends or holidays) needed to complete an activity. It gives a realistic view of how long an activity is scheduled to take for completion.

End activity   An activity with no logical successors.

End event (of a project)   Event with preceding but no succeeding activities. Note: there may be more than one end event.

Environmental factoring   Use of data relating to an external factor (such as the weather) to modify or bias the value of parameters concerned.

Equivalent activity   Activity that is equivalent, in the probabilistic sense, to any combination of series and parallel activities.

Estimate   A quantified assessment of the resources required to complete part or all of a project. The prediction of the quantitative result. It is usually applied to project costs, resources and durations.

Estimate at completion (EAC)   A value expressed in either money and/or hours, to represent the projected final costs of work when completed. The EAC is calculated as ETC + ACWP.

Estimate to complete (ETC)   The value expressed in either money or hours developed to represent the cost of the work required to complete a task.

Estimating The act of combining results of post-project reviews, metrics, consultation and informed assessment to arrive at time and resource requirements for an activity.

Event State in the progress of a project after the completion of all preceding activities but before the start of any succeeding activity. (A defined point that is the beginning or end of an activity.)

Exception report Focused report drawing attention to instances where planned and actual results are expected to be, or are already, significantly different. Note: an exception report is usually triggered when actual values are expected to cross a predetermined threshold that is set with reference to the project plan. The actual values may be trending better or worse than plan.

Exceptions Occurrences that cause deviation from a plan, such as issues, change requests and risks. Exceptions can also refer to items where the cost variance and schedule variance exceed predefined thresholds.

Exclusive OR relationship Logical relationship indicating that only one of the possible activities can be undertaken.

Execution phase The phase of a project in which work towards direct achievement of the project's objectives and the production of the project's deliverables occurs. Sometimes called the implementation phase.

Expenditure A charge against available funds, evidenced by a voucher, claim or other documents. Expenditures represent the actual payment of funds.

External constraint A constraint from outside the project network.

Fallback plan A plan for an alternative course of action that can be adopted to overcome the consequences of a risk, should it occur (including carrying out any advance activities that may be required to render the plan practical).

Fast-tracking Reducing the duration of a project usually by overlapping phases or activities that were originally planned to be done sequentially. (The process of reducing the number of sequential relationships and replacing them typically with parallel relationships, usually to achieve shorter overall durations but often with increased risk.)

Feasibility phase   The project phase that demonstrates that the client's requirement can be achieved and identifies and evaluates the options to determine the one preferred solution.

Feasibility study   Analysis to determine whether a course of action is possible within the terms of reference of the project.

Feasible schedule   Any schedule capable of implementation within the externally determined constraints of time and/or resource limits.

Final report   Post-implementation report. Normally a retrospective report that formally closes the project, having handed over the project deliverables for operational use. Note: the report should draw attention to experiences that may be of benefit to future projects and may form part of the accountability of the project team.

Finish date   The actual or estimated time associated with an activity's completion.

Finishing activity   The last activity that must be completed before a project can be considered finished. This activity is not a predecessor to any other activity – it has no successors.

Finish-to-finish lag   The minimum amount of time that must pass between the finish of one activity and the finish of its successor(s).

Finish-to-start lag   The minimum amount of time that must pass between the finish of one activity and the start of its successor(s). The default finish-to-start lag is zero.

Fix fixed-price contract   A contract where the buyer pays a set amount to the seller regardless of that seller's cost to complete the contract.

Fixed date   A calendar date (associated with a plan) that cannot be moved or changed during the schedule.

Fixed-duration scheduling A scheduling method in which, regardless of the number of resources assigned to the task, the duration remains the same.

Fixed finish See 'imposed finish'.

Fixed-price contracts A generic category of contracts based on the establishment of firm legal commitments to complete the required work. A performing contractor is legally obligated to finish the job, no matter how much it costs to do so.

Fixed start See 'imposed start'.

Float Time available for an activity or path in addition to its planned duration. (Float is the amount of time that an activity can slip past its earliest completion date without delaying the rest of the project. The calculation depends on the float type. See 'free float', 'positive float' and 'total float'.

Forecast at completion Scheduled cost for a task.

Forecast final cost See 'estimate at completion'.

Forward pass A procedure whereby the earliest event times or the earliest start and finish times for the activities of a network are calculated.

Free float Time by which an activity may be delayed or extended without affecting the start of any succeeding activity. Note: free float can never be negative.

Functional organisation Management structure where specific functions of an organisation are grouped into specialist departments providing dedicated services. Note: examples of functional organisations are finance, marketing and design departments.

Funding profile An estimate of funding requirements over time.

Gantt chart Particular type of bar chart showing planned activity against time. Note: 'Gantt chart', although named for a particular type of bar chart, is in current usage as a name for bar charts in general. (A Gantt chart is a time-phased graphic display of activity durations. Activities are listed with other tabular information on the left side with time intervals over the bars. Activity durations are shown in the form of horizontal bars.)

Goal A one-sentence definition of specifically what will be accomplished, while incorporating an event signifying completion.

Hammock Activity joining two specified points that span two or more activities. Note: its duration is initially unspecified and is determined only by the durations of the specified activities. Note: hammocks are usually used to collect time-dependent information, e.g. overheads. (A group of activities, milestones, or other hammocks aggregated together for analysis or reporting purposes. Sometimes used to describe an activity such as management support that has no duration of its own but derives one from the time difference between the two points to which it is connected).

Hand-over The formal process of transferring responsibility for and ownership of the products of a project to the operator or owner.

Hierarchical coding structure A coding system that can be represented as a multilevel tree structure in which every code except those at the top of the tree has a parent code.

Hierarchy of networks Range of networks at different levels of detail, from summary down to working levels, showing the relationships between those networks.

Histogram A graphic display of planned and/or actual resource usage over a period of time. It is in the form of a vertical bar chart, the height of each bar representing the quantity of resource usage in a given time unit. Bars may be single, multiple or show stacked resources.

Holiday An otherwise valid working day that has been designated as exempt from work.

Host organisation Organisation that provides the administrative and logistical support for the project.

Hypercritical activities Activities on the critical path with negative float.

Impact   The assessment of the adverse effects of an occurring risk.

Impact analysis   Assessing the merits of pursuing a particular course of action.

Implementation phase   The project phase that develops the chosen solution into a completed deliverable. (Note: realisation is the internationally accepted and preferred term for implementation.)

Imposed date   Point in time determined by circumstances outside the network. Note: a symbol is inserted immediately above the event concerned on activity-on-arrow networks or adjacent and connected to the appropriate corner of the node on activity-on-node networks.

Imposed finish   A finish date imposed on an activity by external constraints.

Imposed start   A start date imposed on an activity by external constraints.

Inclusive OR relationship   Logical relationship indicating that at least one, but not necessarily all, of the activities have to be undertaken.

INCOTERMS   A set of international terms defining conditions for delivery and shipping of equipment and materials.

Incurred costs   Sum of actual and committed costs, whether invoiced/paid or not, at a specified time.

Indirect cost   Costs associated with a project that cannot be directly attributed to an activity or group of activities. (Resources expended which are not directly identified to any specific contract, project, product or service, such as overheads and general administration.)

In-house project   A project commissioned and carried out entirely within a single organisation.

Initiation   Committing the organisation to begin a project.

In progress An activity that has been started but not yet completed.

Integrated logistics support Disciplined approach to activities necessary to cause support considerations to be integrated into product design and to develop support arrangements that are consistently related to design and to each other and provide the necessary support at the beginning and during customer use at optimum cost.

Integration The process of bringing together people, activities and other things to perform effectively.

Internal rate of return (IRR) Discount rate at which the net present value of a future cash flow is zero. Note: IRR is a special case of the 'discounted cash flow' procedures.

Inverted matrix A project-oriented organisation structure that employs permanent specialists to support projects.

Issue An immediate problem requiring resolution.

Key events Major events, the achievement of which is deemed to be critical to the execution of the project.

Key performance indicators Measurable indicators that will be used to report progress that is chosen to reflect the critical success factors of the project.

Labour rate variances Differences between planned and actual labour rates.

Ladder Device for representing a set of overlapping activities in a network diagram. Note: the start and finish of each succeeding activity is linked only to the start and finish of the preceding activity by lead and lag activities, which consume only time.

Lag In a network diagram, the minimum necessary lapse of time between the finish of one activity and the finish of an overlapping activity. Also, the delay incurred between two specified activities.

Late dates Calculated in the backward pass of time analysis, late dates are the latest dates by which an activity can be allowed to start or finish.

Late event date Calculated from backward pass, it is the latest date an event can occur.

Latest event time Latest time by which an event has to occur within the logical and imposed constraints of the network, without affecting the total project duration.

Latest finish time The latest possible time by which an activity has to finish within the logical activity and imposed constraints of the network, without affecting the total project duration.

Lead In a network diagram, the minimum necessary lapse of time between the start of one activity and the start of an overlapping activity.

Lead contractor The contractor who has responsibility for overall project management and quality assurance.

Leadership Getting others to follow.

Letter of intent A letter indicating an intent to sign a contract, usually so that work can commence prior to signing that contract.

Levelling See 'resource levelling'.

Life-cycle A sequence of defined stages over the full duration of a project.

Life-cycle costing When evaluating alternatives, life-cycle costing is the concept of including acquisition, operating and disposal costs.

Likelihood Assessment of the probability that a risk will occur.

Line manager The manager of any group that makes a product or performs a service.

Linked bar chart A bar chart that shows the dependency links between activities.

Logic   See 'network logic'.

Logic diagram   A diagram that displays the logical relationships between project activities.

Logical relationship   A logical relationship is based on the dependency between two project activities or between a project activity and a milestone.

Loop   An error in a network which results in a later activity imposing a logical restraint on an earlier activity.

Management by project   A term used to describe normal management processes that are being project managed.

Management development   All aspects of staff planning, recruitment, development, training and assessment.

Management reserve   A central contingency pool. Sum of money held as an overall contingency to cover the cost impact of some unexpected event occurring. Note: this is self-insurance.

Master network   Network showing the complete project, from which more detailed networks are derived.

Master schedule   A high-level summary project schedule that identifies major activities and milestones.

Material   Property which may be incorporated into or attached to an end item to be delivered under a contract or which may be consumed or expended in the performance of a contract. It includes, but is not limited to, raw and processed material, parts, components, assemblies, fuels and lubricants, and small tools and supplies which may be consumed in normal use in the performance of a contract.

Matrix organisation   An organisational structure where the project manager and the functional managers share the responsibility of assigning priorities and for directing the work.

Methodology   A documented process for the management of projects that contains procedures, definitions and roles and responsibilities.

Mid-stage assessment   An assessment in the middle of a project that can be held for several reasons: 1) at the request of the project board; 2) to authorise work on the next stage before the current one is completed; 3) to allow for a formal review in the middle of a long project; or 4) to review exception plans.

Milestone   A key event. An event selected for its importance in the project. Note: milestones are commonly used in relation to progress. (A milestone is often chosen to represent the start of a new phase or completion of a major deliverable. They are used to monitor progress at summary level. Milestones are activities of zero duration.)

Milestone plan   A plan containing only milestones which highlight key points of the project.

Milestone schedule   A schedule that identifies the major milestones. See also 'master schedule'.

Mission statement   Brief summary, approximately one or two sentences, that sums up the background, purposes and benefits of the project.

Mitigation   Working to reduce risk by lowering its chances of occurring or by reducing its effect if it occurs.

Mobilisation   The bringing together of project personnel and securing equipment and facilities. Carried out during project start-up phases.

Monitoring   The recording, analysing and reporting of project performance as compared with the plan.

Monte Carlo simulation   A technique used to estimate the likely range of outcomes from a complex process by simulating the process under randomly selected conditions a large number of times.

Multi-project   A project consisting of multiple subprojects.

Multi-project analysis   Multi-project analysis is used to analyse the impact and interaction of activities and resources whose progress affects the progress of a group of projects or for projects with shared resources or both. Multi-project analysis can also be used for composite reporting on projects having no dependencies or resources in common.

Multi-project management   Managing multiple projects that are interconnected either logically or by shared resources.

Multi-project scheduling   Use of the techniques of resource allocation to schedule more than one project concurrently.

Near-critical activity   A low total float activity.

Negative total float   Time by which the duration of an activity of path has to be reduced in order to permit a limiting imposed date to be achieved.

Negotiated contract cost   The estimated cost negotiated in a cost-plus-fixed-fee contract or the negotiated contract target cost in either a fixed-price-incentive contract or a cost-plus-incentive-fee contract. See also 'contract target cost'.

Negotiation   The art of satisfying needs by reaching agreement or compromise with other parties.

Net present value   Aggregate of future net cash flows discounted back to a common base date, usually the present.

Network   A pictorial presentation of project data in which the project logic is the main determinant of the placements of the activities in the drawing. Frequently called a flowchart, PERT chart, logic drawing or logic diagram.

Network analysis   Method used for calculating a project's critical path and activity times and floats. See also 'critical path analysis', 'project network techniques'.

Network interface   Activity or event common to two or more network diagrams.

Network logic   The collection of activity dependencies that make up a project network.

Nodes   Points in a network at which arrows start and finish.

Non-recurring costs   Expenditures against specific tasks that are expected to occur only once on a given project.

Non-splittable activity   An activity that, once started, has to be completed to plan without interruption. Note: resources should not be diverted from a non-splittable activity to another activity.

Not earlier than   A restriction on an activity that indicates it may not start or end earlier than a specified date.

Not later than   A restriction on an activity that indicates it may not start or end later than a specified date.

Objectives   Predetermined results towards which effort is directed.

Operation phase   Period when the completed deliverable is used and maintained in service for its intended purpose.

Opportunity   The opposite of a risk. The chance to enhance the project benefits.

Order of magnitude estimate   An estimate carried out to give very approximate indication of likely out-turn costs.

Organisation design   The design of the most appropriate organisational design for a project.

Organisational breakdown structure (OBS)   Hierarchical way in which the organisation may be divided into management levels and groups, for planning and control purposes.

Organisational planning   The process of identifying, assigning and documenting project responsibilities and relationships.

Original budget   The initial budget established at or near the time a contract was signed or a project authorised, based on the negotiated contract cost or management's authorisation.

Original duration   The duration of activities or groups of activities as recorded in the baseline schedule.

Other direct costs (ODC)   A group of accounting elements which can be isolated to specific tasks, other than labour and material. Included in the ODC are such items as travel, computer time and services.

Out-of-sequence progress   Progress that has been reported even though activities that have been deemed predecessors in project logic have not been completed.

Output format   Information that governs the final appearance of a report or drawing (usually refers to computer-generated documents).

Outsourcing   Contracting out, buying in facilities or work (as opposed to using in-house resources).

Overall change control   Co-ordinating and controlling changes across an entire project.

Overhead   Costs incurred in the operation of a business that cannot be directly related to the individual products or services being produced. See also 'indirect cost'.

Overrun   Costs incurred in excess of the contract target costs on an incentive-type contract or the estimated costs on a fixed-fee contract. An overrun is that value of costs which are needed to complete a project, over that value originally authorised by management.

Parallel activities   Two or more activities that can be done at the same time. This allows a project to be completed faster than if the activities were arranged serially.

Parent activity   Task within the work breakdown structure that embodies several subordinate 'child' tasks.

Parties (to a contract)   The people or companies which sign a contract with one another.

Path   Activity or an unbroken sequence of activities in a project network (Refer to 'critical path method' for information on critical and non-critical paths.)

Percent complete   A measure of the completion status of a partially completed activity. May be aggregated to sections of a project or the whole project.

Performance measurement techniques   The methods used to estimate earned value. Different methods are appropriate to different work packages, either due to the nature of the work or to the planned duration of the work package.

Performance specification   Statement of the totality of needs expressed by the benefits, features, characteristics, process conditions, boundaries and constraints that together define the expected performance of a deliverable. Note: a performance specification should provide for innovation and alternative solutions by not defining or unduly constraining the technical attributes of the intended deliverable.

Performing   A teambuilding stage where the emphasis is on the work currently being performed.

Phase (of a project)   That part of a project during which a set of related and interlinked activities are performed. Note: a project consists of a series of phases that together constitute the whole project life-cycle.

Physical percent complete   The percentage of the work content of an activity that has been achieved.

Pilot   A form of testing a new development and its implementation prior to committing to its full release.

Plan   An intended future course of action. It is owned by the project manager, it is the basis of the project controls and includes the 'what', the 'how', the 'when' and the 'who'.

Planned activity   An activity not yet started.

Planned cost   Estimated cost of achieving a specified objective.

Planner   A member of a project team or project support office with the responsibility for planning, scheduling and tracking of projects. They are often primarily concerned with schedule, progress and manpower resources.

Planning The process of identifying the means, resources and actions necessary to accomplish an objective.

Planning stage The stage prior to the implementation stage when product activity, resource and quality plans are produced.

Portfolio A grouping or bundle of projects, collected together for management convenience. They may or may not have a common objective – they are often related only by the use of common resources.

Portfolio management The management of a number of projects that do not share a common objective.

Positive float Defined as the amount of time that an activity's start can be delayed without affecting the project completion date. An activity with positive float is not on the critical path and is called a non-critical activity. The difference between early and late dates (start or finish) determines the amount of float.

Post-implementation review A review between 6–12 months after a system in a project has met its objectives to verify that it continues to meet user requirements.

Post-project appraisal An evaluation that provides feedback in order to learn for the future.

Precedence diagram method One of two methods of representing project as networks, in which the activities are represented by nodes and the relationships between them by arrows.

Precedence network A multiple dependency network. An activity-on-node network in which a sequence arrow represents one of four forms of precedence relationship, depending on the positioning of the head and the tail of the sequence arrow. The relationships are:

- Start of activity depends on finish of preceding activity, either immediately or after a lapse of time.

- Finish of activity depends on finish of preceding activity, either immediately or after a lapse of time.

- Start of activity depends on start of preceding activity, either immediately or after a lapse of time.

- Finish of activity depends on start of preceding activity, either immediately or after a lapse of time.

Preceding event   In an activity-on-arrow network, an event at the beginning of an activity.

Pre-commissioning   That work which is carried out prior to commissioning in order to demonstrate that commissioning may be safely undertaken.

Predecessor   An activity that must be completed (or be partially completed) before a specified activity can begin.

Predecessor activity   In the precedence diagramming method this is an activity which logically precedes the current activity.

Prime or lead contractor   A main supplier who has a contract for much or all of the work on a project.

Probabilistic network   Network containing alternative paths with which probabilities are associated.

Probability   Likelihood of a risk occurring.

Process   Set of inter-related resources and activities which transform inputs into outputs.

Procurement   The securing of goods or services.

Procurement planning   Determining what to procure and when.

Product breakdown structure   A hierarchy of deliverable products which are required to be produced on the project. It forms the base document from which the execution strategy and product-based work breakdown structure may be derived. It provides a guide for configuration control.

Product description   The description of the purpose form and components of a product. It should always be used as a basis for acceptance of the product by the customer.

Product flow diagram   Represents how the products are produced by identifying their derivation and the dependencies between them.

Programme   A broad effort encompassing a number of projects and/or functional activities with a common purpose.

Programme benefits review   A review to assess whether targets have been reached and to measure the performance levels in the resulting business operations.

Programme director   The senior manager with the responsibility for the overall success of the programme.

Programme directorate   A committee that directs the programme when circumstances arise where there is no individual to direct the programme.

Programme evaluation and review technique (PERT)   Also known as performance evaluation and review technique, and probability evaluation review technique. A project management technique for determining how much time a project needs before it is completed. Each activity is assigned a best, worst and most probable completion time estimate. These estimates are used to determine the average completion time. The average times are used to calculate the critical path and the standard deviation of completion times for the entire project.

Programme management   The effective management of several individual but related projects or functional activities in order to produce an overall system that works effectively.

Programme management office   The office responsible for the business and technical management of a specific content or programme.

Programme manager   Individual or body with responsibility for managing a group of projects.

Programme support office   A group that gives administrative support to the programme manager and the programme executive.

Progress   The partial completion of a project or a measure of the same.

Progress payments   Payments made to a contractor during the life of a fixed-price type contract, on the basis of some agreed-to formula, for example, budget cost of work performed or simply costs incurred.

Progress report   A regular report to senior personnel, sponsors or stakeholders summarising the progress of a project, including key events, milestones, costs and other issues.

Project   Unique set of co-ordinated activities, with definite starting and finishing points, undertaken by an individual or organisation to meet specific objectives within defined time, cost and performance parameters. (See also BS ISO 10006). (Alternative definition: an endeavour in which human, material and financial resources are organised in a novel way to deliver a unique scope of work of given specification, often within constraints of cost and time, and to achieve beneficial change defined by quantitative and qualitative objectives.)

Project appraisal   The discipline of calculating the viability of a project.

Project base date   Reference date used as a basis for the start of a project calendar.

Project board   The body to which the project manager is accountable for achieving the project objectives.

Project brief   Statement that describes the purpose, cost, time and performance requirements/constraints for a project. (A statement of reference terms for a project. A written statement of the client's goals and requirements in relation to the project.)

Project calendar   A calendar that defines global project working and non-working periods.

Project champion   Person within the parent organisation who promotes and defends a project.

Project closure   Formal termination of a project at any point during its life.

Project co-ordination   Communication linking various areas of a project to ensure the transfer of information or hardware at interface points at the appropriate times and identification of any further necessary resources.

Project co-ordination procedure   Defines the parties relevant to the project and the approved means of communicating between them.

Project cost management   A subset of project management that includes resource planning, cost estimating, cost control and cost budgeting in an effort to complete the project within its approved budget.

Project culture   The general attitude toward projects within the business.

Project definition   A report that defines a project, i.e. why it is required, what will be done, how, when and where it will be delivered, the organisation and resources required, the standards and procedures to be followed.

Project director   The manager of a very large project that demands senior-level responsibility or the person at the board level in an organisation who has the overall responsibility for the project's management.

Project environment   The context within which the project is formulated, assessed and realised. This includes all external factors that have an impact on the project.

Project evaluation   A documented review of the project's performance, produced at project closure. It ensures that the experience of the project is recorded for the benefit of others.

Project file   A file containing the overall plans of a project and any other important documents.

Project initiation   The beginning of a project at which point certain management activities are required to ensure that the project is established with clear reference terms and adequate management structure.

Project initiation document   A document approved by the project board at project initiation that defines the terms of reference for the project.

Project issue report   A report that raises either technical or managerial issues in a project.

Project life-cycle   All phases or stages between a project's conception and its termination. Note: the project life-cycle may include the operation and disposal of project deliverables. This is usually known as an 'extended life cycle'.

Project life-cycle cost   Cumulative cost of a project over its whole life cycle.

Project log   A project diary. A chronological record of all significant occurrences throughout the project.

Project logic   The relationships between the various activities in a project.

Project logic drawing   A representation of the logical relationships of a project.

Project management   Planning, monitoring and control of all aspects of a project and the motivation of all those involved in it to achieve the project objectives on time and to the specified cost, quality and performance. (Alternative definition: the controlled implementation of defined change.)

Project management body of knowledge   An inclusive term that describes the sum of knowledge within the profession of project management. As with other professions, such as law and medicine, the body of knowledge rests with the practitioners and academics who apply and advance it.

Project management plan   A plan prepared by or for the project manager for carrying out a project to meet specific objectives.

Project management software   Computer application software designed to help with planning and controlling resources, costs and schedules of a project. It may also provide facilities for documentation management, risk analysis, etc.

Project management team   Members of the project team who are directly involved in its management.

Project manager   Individual or body with authority, accountability and responsibility for managing a project to achieve specific objectives.

Project matrix   An organisation matrix that is project based in which the functional structures are duplicated in each project.

Project monitoring   Comparison of current project status with what was planned to be done to identify and report any deviations.

Project network   Representation of activities and/or events with their inter-relationships and dependencies.

Project network techniques   Group of techniques that, for the description, analysis, planning and control of a project, considers the logical inter-relationships of all project activities. The group includes techniques concerned with time, resources, costs and other influencing factors, e.g. uncertainty. Note: the terms 'programme evaluation and review technique' (PERT), 'critical path analysis' (CPA), 'critical path method' (CPM) and 'precedence method' refer to particular techniques and should not be used as synonyms for project network.

Project organisation   Structure that is created or evolves to serve the project and its participants. (A term which refers to the structure, roles and responsibilities of the project team and its interfaces with the outside world.)

Project phase   A group of related project activities that come together with the completion of a deliverable.

Project plan   A document for management purposes that gives the basics of a project in terms of its objectives, justification and how the objectives are to be achieved. This document is used as a record of decisions and a means of communication among stakeholders. It gives the supporting detail to the project definition which details the schedule, resource and costs for the project.

Project planning   Developing and maintaining a project plan.

Project portfolio   The constituent parts within a programme.

Project procedures material   A collected set of the management and administrative procedures needed for the project.

Project procurement management   A subset of project management that includes procurement planning, source selection, inquiry, tender assessment, placement of purchase orders and contracts for goods and services, contract and purchase order administration and close-out in an effort to obtain goods and services from outside organisations.

Project progress report   Formal statement that compares the project progress, achievements and expectations with the project plan.

Project quality management   A subset of project management that includes quality planning, quality assurance and quality control to satisfy the needs and purpose of the project.

Project review calendar   Calendar of project review dates, meetings and issues of reports set against project week numbers or dates.

Project risk management   A subset of project management that includes risk identification, risk quantification, risk response development and risk response control in an effort to identify, analyse and respond to project risks.

Project schedule   Project programme (planned dates for starting and completing activities and milestones).

Project scope management   A subset of project management that includes initiation, scope planning, scope definition, scope verification and scope change control in an effort to ensure that the project has all of the necessary work required to complete it.

Project sponsor   The individual or body for whom the project is undertaken, the primary risk taker. The individual representing the sponsoring body and to whom the project manager reports. A person or organisation providing funds for the project.

Project start-up   The creation of the project team.

Project status report   A report on the status of accomplishments and any variances to spending and schedule plans.

Project strategy   A comprehensive definition of how a project will be developed and managed.

Project success/failure criteria   The criteria by which the success or failure of a project may be judged.

Project support office   The central location of planning and project support functions. Often provides personnel and facilities for centralised planning, cost management, estimating, documentation control and sometimes procurement to a number of projects.

Project team   Set of individuals, groups and/or organisations that are responsible to the project manager for undertaking project tasks. (Includes all contractors and consultants.)

Project technical plan   A plan produced at the beginning of a project that addresses technical issues and strategic issues related to quality control and configuration management.

Project time management   A subset of project management that includes activity definition, activity sequencing, activity duration estimating, schedule development and schedule control in order to complete the project on time.

Public relations   An activity meant to improve the project organisation's environment in order to improve project performance and reception.

Qualitative risk analysis   A generic term for subjective methods of assessing risks.

Quality   A trait or characteristic used to measure the degree of excellence of a product or service. Meeting customers' needs.

Quality assurance (QA)   The process of evaluating overall project performance on a regular basis to provide confidence that the project will satisfy the relevant quality standards.

Quality assurance plan   A plan that guarantees a quality approach and conformance to all customer requirements for all activities in a project.

Quality audit   An official examination to determine whether practices conform to specified standards or a critical analysis of whether a deliverable meets quality criteria.

Quality control (QC)   The process of monitoring specific project results to determine whether they comply with relevant standards and identifying ways to eliminate causes of unsatisfactory performance.

Quality criteria   The characteristics of a product that determine whether it meets certain requirements.

Quality guide   Describes quality and configuration management procedures and is aimed at people directly involved with quality reviews, configuration management and technical exceptions.

Quality plan (for a project)   That part of the project plan that concerns quality management and quality assurance strategies. (See also ISO 10006.)

Quality planning   Determining which quality standards are necessary and how to apply them.

Quality review   A review of a product against an established set of quality criteria.

Recurring costs   Expenditures against specific tasks that would occur on a repetitive basis. Examples are hire of computer equipment, tool maintenance, etc.

Relationship   A logical connection between two activities.

Remaining duration   Time needed to complete the remainder of an activity or project.

Re-planning   Actions performed for any remaining effort within project scope. Often the cost and/or schedule variances are zeroed out at this time for history items.

Request for change   A proposal by the project manager for a change to the project as a result of a project issue report.

Request for proposal   A bid document used to request proposals from prospective sellers of products or services.

Request for quotation   Equivalent to a request for proposal but with more specific application areas.

Requirements   A negotiated set of measurable customer wants and needs.

Requirements definition   Statement of the needs that a project has to satisfy.

Resource   Any variable capable of definition that is required for the completion of an activity and may constrain the project.

Note 1: a resource may be non-storable so that its availability has to be renewed for each time period (even if it was not utilised in previous time periods).

Note 2: a resource may be storable so that it remains available unless depleted by usage. Such a resource may also be replenished by activities producing credited and storable resource. (Resources can be people, equipment, facilities, funding or anything else needed to perform the work of a project.)

Resource aggregation   Summation of the requirements for each resource and for each time period. Note: where the earliest start time of an activity is used alone, it is often termed an 'early start' aggregation. Similarly a 'late start' aggregation uses the latest start times.

Resource allocation   Scheduling of activities and the resources required by those activities, so that predetermined constraints of resource availability and/or project time are not exceeded.

Resource analysis   The process of analysing and optimising the use of resources on a project. Often uses resource levelling and resource smoothing techniques.

Resource assignment   The work on an activity related to a specific resource.

Resource availability   The level of availability of a resource, which may vary over time.

Resource breakdown structure (RBS)   A hierarchical structure of resources that enables scheduling at the detailed requirements level and roll-up of both requirements and availabilities to a higher level.

Resource calendar   A calendar that defines the working and non-working patterns for specific resources.

Resource constraint   Limitation due to the availability of a resource.

Resource cumulation   Process of accumulating the requirements for each resource to give the total required to date at all times throughout the project.

Resource-driven task durations   Task durations that are driven by the need for scarce resources.

Resource histogram   A view of project data in which resource requirements, usage and availability are shown up using vertical bars against a horizontal time scale.

Resource level   A specified level of resource units required by an activity per time unit.

Resource levelling   See 'resource limited scheduling'.

Resource limited scheduling   Scheduling of activities so that predetermined resource levels are never exceeded. Note: this may cause the minimum overall or specified project duration to be exceeded.

Resource optimisation   A term for resource levelling and resource smoothing.

Resource plan   Part of the definition statement stating how the programme will be resource loaded and what supporting services, infrastructure and third-party services are required.

Resource planning   Evaluating what resources are needed to complete a project and determining the quantity needed.

Resource requirement   The requirement for a particular resource by a particular activity.

Resource scheduling   The process of determining dates on which activities should be performed in order to smooth the demand for resources or to avoid exceeding stated constraints on these restraints.

Resource smoothing   Scheduling of activities within the limits of their float, so that fluctuations in individual resource requirements are minimised. (In smoothing, as opposed to resource levelling, the project completion date may not be delayed.)

Responsibility matrix   A document correlating the work required by a work breakdown structure element to the functional organisations responsible for accomplishing the assigned tasks.

Responsible organisation   A defined unit within the organisation structure which is assigned responsibility for accomplishing specific tasks or cost accounts.

Retention   Part of a payment withheld until the project is completed in order to ensure satisfactory performance or completion of contract terms.

Revenue cost   Expenditure charged to the profit and loss account as incurred or accrued due.

Risk   Combination of the probability or frequency of occurrence of a defined threat or opportunity and the magnitude of the consequences of the occurrence. Note: combination of the likelihood of occurrence of a specified event and its consequences.

Risk analysis   Systematic use of available information to determine how often specified events may occur and the magnitude of their likely consequences.

Risk assessment   The process of identifying potential risks, quantifying their likelihood of occurrence and assessing their likely impact on the project.

Risk avoidance   Planning activities to avoid risks that have been identified.

Risk evaluation   Process used to determine risk management priorities.

Risk event   A discrete occurrence that affects a project.

Risk identification   Process of determining what could pose a risk.

Risk management   Systematic application of policies, procedures, methods and practices to the tasks of identifying, analysing, evaluating, treating and monitoring risk. (The process whereby decisions are made to accept known or assessed risks and/or the implementation of actions to reduce the consequences of probability of occurrence.)

Risk management plan   A document defining how project risk analysis and management are to be implemented in the context of a particular project.

Risk matrix   A matrix with risks located in rows, and with impact and likelihood in columns.

Risk prioritising   Ordering of risks according first to their risk value and then by which risks need to be considered for risk reduction, risk avoidance and risk transfer.

Risk quantification   Process of applying values to the various aspects of a risk (Evaluating the probability of risk event effect and occurrence.)

Risk ranking   Allocating a classification to the impact and likelihood of a risk.

Risk reduction   Action taken to reduce the likelihood and impact of a risk.

Risk register   Formal record of identified risks. (A body of information listing all the risks identified for the project, explaining the nature of each risk and recording information relevant to its assessment and management.)

Risk response   Contingency plans to manage a risk should it materialise. (Action to reduce the probability of the risk arising or to reduce the significance of its detrimental impact if it does arise.)

Risk, secondary   Risk that can occur as a result of treating a risk.

Risk sharing   Diminution of a risk by sharing it with others, usually for some consideration.

Risk transfer   A contractual arrangement between two parties for delivery and acceptance of a product where the liability for the costs of a risk is transferred from one party to another.

Risk treatment   Selection and implementation of appropriate options for dealing with risk.

Safety plan   The standards and methods which minimise to an acceptable level the likelihood of accident or damage to people or equipment.

S-curve   A display of cumulative costs, labour hours or other quantities plotted against time.

Schedule   The timetable for a project. It shows how project tasks and milestones are planned out over a period of time.

Schedule control   Controlling schedule changes.

Schedule dates   Start and finish dates calculated with regard to resource or external constraints as well as project logic.

Schedule performance index (SPI)   Ratio of work accomplished versus work planned, for a specified time period. The SPI is an efficiency rating for work accomplishment, comparing work accomplished with what should have been accomplished.

Schedule variance (cost)   The difference between the budgeted cost of work performed and the budgeted cost of work scheduled at any point in time.

Scheduled finish   The earliest date on which an activity can finish, having regard to resource or external constraints as well as project logic.

Scheduled start   The earliest date on which an activity can start, having regard to resource or external constraints as well as project logic.

Scheduling   The process of determining when project activities will take place depending on defined durations and precedent activities. Schedule constraints specify when an activity should start or end based on duration, predecessors, external predecessor relationships, resource availability or target dates.

Scope   The sum of work content of a project.

Scope change   Any change in a project scope that requires a change in the project's cost or schedule.

Scope change control   Controlling changes to the scope.

Scope of work   A description of the work to be accomplished or resources to be supplied.

Scope verification   Ensuring all identified project deliverables have been completed satisfactorily.

Secondary risk   The risk that may occur as a result of invoking a risk response of fallback plan.

Secondment matrix   An organisational structure whereby team members are seconded from their respective departments to the project and are responsible to the project manager.

Sequence   The order in which activities will occur with respect to one another.

Slack   Calculated time span within which an event has to occur within the logical and imposed constraints of the network, without affecting the total project duration.

Note 1: it may be made negative by an imposed date.

Note 2: the term 'slack' is used as referring only to an event.

Slip chart   A pictorial representation of the predicted completion dates of milestones (also referred to as trend chart).

Slippage   The amount of slack or float time used up by the current activity due to a delayed start or increased duration.

Soft project   A project that is intended to bring about change and does not have a physical end product.

Soft skills   Include teambuilding, conflict management and negotiation.

Source selection   Choosing from potential contractors.

Splittable activity   Activity that can be interrupted in order to allow its resources to be transferred temporarily to another activity.

Sponsor   Individual or body for whom the project is undertaken and who is the primary risk taker.

Stage   A natural high-level subsection of a project that has its own organisational structure, life span and manager.

Stage payment   Payment part way through a project at some predetermined milestone.

Stakeholder   A person or group of people who have a vested interest in the success of an organisation and the environment in which the organisation operates. (Project stakeholders are people or organisations with a vested interest in the environment, performance and/or outcome of the project.)

Start event of a project   Event with succeeding but no preceding activities. Note: there may be more than one start event.

Starting activity   A starting activity has no predecessors. It does not have to wait for any other activity to start.

Start-to-start lag   The minimum amount of time that must pass between the start of one activity and the start of its successor(s). This may be expressed in terms of duration or percentage.

Statement of work   A document stating the requirements for a given project task.

Status reports   Written reports given to both the project team and to a responsible person on a regular basis stating the status of an activity, work package or whole project. Status reports should be used to control the project and to keep management informed of project status.

Steering group   A body established to monitor the project and give guidance to the project sponsor or project manager.

Subcontract   A contractual document which legally transfers the responsibility and effort of providing goods, services, data or other hardware from one firm to another.

Subcontractor   An organisation that supplies goods or services to a supplier.

Subnet or subnetwork   A division of a project network diagram representing a subproject.

Subproject   A group of activities represented as a single activity in a higher level of the same.

Success criteria   Criteria to be used for judging whether the project is successful.

Success factors   Critical factors that will ensure achievement of success criteria.

Successor   An activity whose start or finish depends on the start or finish of a predecessor activity.

Sunk costs   Unavoidable costs (even if the project were to be terminated).

Super-critical activity   An activity that is behind schedule is considered to be super-critical if it has been delayed to a point where its float is calculated to be a negative value.

Supplier   Includes contractors, consultants and any organisation that supplies services or goods to the customer.

System   The complete technical output of the project including technical products.

Systems and procedures   Detail the standard methods, practices and procedures of handling frequently occurring events within the project.

Systems management   Management that includes the prime activities of systems analysis, systems design and systems development.

Target completion date   A date which contractors strive towards for completion of the activity.

Target date   Date imposed on activity or project by the user. There are two types of target dates: target start dates and target finish dates.

Target finish – activity   Target finish is the user's imposed finish date for an activity. A target finish date is used if there are predefined commitment dates.

Target finish date   The date planned to finish work on an activity.

Target finish – project   A user's target finish date can be imposed on a project as a whole. A target finish date is used if there is a predefined completion date.

Target start – activity   Target start is an imposed starting date on an activity.

Target start date   The date planned to start work on an activity.

Task   The smallest indivisible part of an activity when it is broken down to a level best understood and performed by a specific person or organisation.

Team   A team is made up of two or more people working interdependently towards a common goal and a shared reward.

Teambuilding   The ability to gather the right people to join a project team and get them working together for the benefit of a project.

Team development   Developing skills, as a group and individually, that enhance project performance.

Team leader   Person responsible for leading a team.

Technical assurance   The monitoring of the technical integrity of products.

Technical guide   A document that guides managers, team leaders and technical assurance co-ordinators on planning the production of products.

Technical products   Products produced by a project for an end-user.

Tender   A document proposing to meet a specification in a certain way and at a stated price (or on a particular financial basis), an offer of price and conditions under which the tenderer is willing to undertake the work for the client.

Termination   Completion of the project, upon formal acceptance of its deliverables by the client and/or the disposal of such deliverables at the end of their life.

Terms of reference   A specification of a team member's responsibilities and authorities within the project.

Tied activities   Activities that have to be performed sequentially or within a predetermined time of each other.

Time analysis   The process of calculating the early and late dates for each activity on a project, based on the duration of the activities and the logical relations between them.

Time-based network   A linked bar chart, a bar chart that shows the logical and time relationships between the activities.

Time-limited resource scheduling The production of scheduled dates in which resource constraints may be relaxed in order to avoid any delay in project completion.

Time-limited scheduling Scheduling of activities so that the specified project duration or any imposed dates are not exceeded. Note: this may cause the envisaged resource levels to be exceeded.

Time now Specified date from which the forward analysis is deemed to commence. (The date to which current progress is reported. Sometimes referred to as the status date because all progress information entered for a project should be correct as of this date.)

Time recording The recording of effort expended on each activity in order to update a project plan.

Time-scaled logic drawing A drawing that displays the logical connection between activities in the context of a time scale in which each horizontal position represents a point in time.

Time-scaled network diagram A project network diagram drawn so that the positioning of the activity represents scheduled duration.

Time sheet A means of recording the actual effort expended against project and non-project activities.

Top-down cost estimating The total project cost is estimated based on historical costs and other project variables and then subdivided down to individual activities.

Total float Time by which an activity may be delayed or extended without affecting the total project duration (or violating a target finish date).

Total quality management (TQM) A strategic, integrated management system for customer satisfaction that guides all employees in every aspect of their work.

Transit time Dependency link that requires time and no other resources. It may be a negative time.

Turnaround report   A report created especially for the various responsible managers to enter their progress status against a list of activities that are scheduled to be in progress during a particular time window.

Unlimited schedule   Infinite schedule, schedule produced without resource constraint.

Users   The group of people who are intended to benefit from the project.

Value   A standard, principle or quality considered worthwhile or desirable.

Value engineering   A technique for analysing qualitative and quantitative costs and benefits of component parts of a proposed system.

Value management   A structured means of improving business effectiveness that includes the use of management techniques such as value engineering and value analysis.

Value planning   A technique for assessing, before significant investment is made, the desirability of a proposal based on the value that will accrue to the organisation from that proposal.

Variance   A discrepancy between the actual and planned performance on a project, either in terms of schedule or cost.

Variance at completion   The difference between budget at complete and estimate at complete.

Variation   A change in scope or timing or work which a supplier is obliged to do under a contract.

Variation order   The document authorising an approved technical change or variation.

What-if-analysis   The process of evaluating alternative strategies.

What-if-simulation   Changing the value of the parameters of the project network to study its behaviour under various conditions of its operation.

Work   The total number of hours, people or effort required to complete a task.

Work breakdown code   A code that represents the 'family tree' of an element in a work breakdown structure.

Work breakdown structure (WBS)   Way in which a project may be divided by level into discrete groups for programming, cost planning and control purposes. See also 'work package'. (The WBS is a tool for defining the hierarchical breakdown of work required to deliver the products of a project. Major categories are broken down into smaller components. These are subdivided until the lowest required level of detail is established. The lowest units of the WBS become the activities in a project. The WBS defines the total work to be undertaken on the project and provides a structure for all project control systems.)

Workload   Workload is the amount of work units assigned to a resource over a period of time.

Work package   A group of related tasks that are defined at the same level within a work breakdown structure. (In traditional cost/schedule systems, the criteria for defining work packages are as follows: 1) each work package is clearly distinguishable from all other work packages in the programme; 2) each work package has a scheduled start and finish date; 3) each work package has an assigned budget that is time-phased over the duration of the work package; 4) each work package either has a relatively short duration or can be divided into a series of milestones whose status can be objectively measured; 5) each work package has a schedule that is integrated with higher-level schedules.)

Work units   Provide the measurement units for resources. For example, people as a resource can be measured by the number of hours they work.

Zero float   A condition where there is no excess time between activities. An activity with zero float is considered a critical activity.

APPENDIX

2 GLOSSARY OF STRUCTURE AND GENE TERMS

Attractor The means by which a system is bound to a particular behaviour in terms of a single point, a regular cycle or to more complex behaviour (see also strange attractor).

Bounded instability Complex oscillations within a system that do not provoke completely unstable behaviour.

Dependency (normal) The reliance placed by one activity on the completion of one or more immediately preceding activities in a project before it can commence (a finish-start relationship).

Design The making of decisions concerning a structured entity.

Discontinuity The existence of a significant difference between two systems with regard to territory, time, technology.

DNA The means by which genes carry information.

Chromosome A thread-like structure carrying genes.

Entropy The conversion of useful energy into useless (in terms of doing work) energy.

Equifinality Systems reaching the same end point despite different start conditions and paths.

Fractal A shape with geometrical self-similarity at all scales.

Gene A unit within a chromosome that carries hereditary information.

Genome   The haploid set of chromosomes for a specific organism.

Interdependence   The existence of a dependency link or links between one activity in a process chain within a project and one or more other activities, in one or more other process chains in that project and/or in other projects.

Lambda   A measure of the quantity of information within a system environment.

Performance specification   A statement of the totality of needs expressed by the benefits, features, characteristics, process conditions, boundaries and constraints that together define the expected performance of a deliverable.

Phase transition   A sudden change of a qualitative nature in a system's behaviour.

Precedence   The need for certain activities to be completed before other activities within a project.

Strange attractor   One having multiple points of attraction within a finite space and binding a system to unstable behaviour.

REFERENCES

APM (2000) *Project Management Body of Knowledge* (ed. M. Dixon), 4th edn. www.apm.org.uk

Aufrecht, S.E. (2001) Why should a manager cross the road? The appropriate use of humour in public organizations. Proceedings, *Critical Management Studies*, Manchester School of Management, UMIST, Manchester.

Banner, D.K. & Gagne, T.E. (1995) *Designing Effective Organisations*, Sage Publications Inc., Thousand Oaks, California.

Barford, K. & Churchouse, S. (2000) *Flexible Working Practices*, Hewitt Associates, St Albans.

Barrington, K. (1960) *The Development of the Architectural Profession in Britain*, George Allen & Unwin Ltd, London.

Belbin, M.R. (1993) *Team Roles at Work*, Butterworth-Heinemann Ltd, Oxford.

Bolman, L.G. & Deal, T.E. (1995) The organisation as theatre. In: *New Thinking in Organisational Behaviour* (ed. H. Tsoukas), Butterworth-Heinemann, Oxford.

Brodnick, R.J. (2000) *Becoming Nonlinear: An Alternative to Linear Thinking*, www.notes3.nms.unt.edu/infrstrt

Buchanan, D.A. & Huczynski, A.A. (1985) *Organisational Behaviour*, Prentice Hall International, London.

Burke, J. (1978) *Connections*, Macmillan Limited, London.

Cantu, C. (2000) *Virtual Teams*, CSWT Reports, Center for the Study of Work Teams, University of North Texas, www.workteams.unt.edu/reports

Carey, J.L. & Doherty, W.O. (1968) Characteristics of professional organizations. In: *Colleagues in Organization: The Social Construction of Professional Work* (ed. R.L. Blankenship), John Wiley and Sons, New York.

Chiavola, B. (2000) AdCoMS – ESPRIT Project No. 22167, http://www.datamation.co.uk/samples/edn00/e/05sp2

CYTGPDC (1999) *Three Gorges Project*, China Yangtze Three Gorges Project Development Corporation, Yichang, China.

Daft, R.L. (2001) *Organization Theory and Design*, 7th edn. South Western College Publishing, Cincinnati, Ohio.

Dainty, A.R.J. & Moore, D.R. (2001) The performance of integrated d&b project teams in unexpected change event management (ed. A. Akintoye), *ARCOM 2000*, 6–9 September, vol. 1 pp. 281–290, Glasgow.

D'Amore, R., Konchady, M. & Orbst, L. (2000) Knowledge mapping aids discovery of organizational information. In: *The Edge Newsletter*, April, www.mitre.org/pubs/edge

D'Herbemont, O. & Cesar, B. (1998) *Managing Sensitive Projects*, Macmillan Press Ltd, London.

References

Duncan, R. (1971) Characteristics of organisational environments and perceived environmental uncertainty. In: *Administrative Science Quarterly*, 16, pp. 313–327.

ECI (European Construction Institute) (1995) *Total Project Management of Construction Safety, Health and Environment*, 2nd edn. Thomas Telford Services Ltd, London.

Edmonds, E.A.& Moran, Thomas P. (1999) Interactive Systems for Supporting the Emergence of Concepts and Ideas. *Proceedings*, Chi Workshop at http://bashful.lboro.ac.uk/chi-wshop/Proceedings/Edmonds.html

Emmitt, S. (2001) Technological gatekeepers: the management of trade literature by design offices. In: *Engineering, Construction and Architectural Management*, 8 (1), pp. 2–6, Blackwell Science Ltd, Oxford.

Fan, C.N.L., Ho, M.H.C. & Ng, V. (2001) Effect of professional socialization on quantity surveyors' ethical perceptions in Hong Kong. In: *Engineering, Construction and Architectural Management*, 8, 4, pp. 304–312, Blackwell Science Ltd, Oxford.

Farr, K. (2000) Organizational learning and knowledge managers. In: *Work Study*, 49, 1, pp. 14–17, MCB University Press, Bradford.

Field, M. & Keller, L. (1998) *Project Management*, The Open University and Thomson Learning, London.

Finke, R.A. (1989) *Principles of Mental Imagery*, MIT Press, Cambridge, MA. Cited in Edmonds & Moran (1999).

Flexibility (2000) The what, why and how of flexible working practices. *Review of new survey findings*, www.flexibility.co.uk/hewittreport.htm

George, J.A. (1996) Virtual best practice. In: *Teams Magazine*, November, pp. 38–45. Cited in Cantu, C. (2000) *Virtual Teams*, CSWT Reports, Center for the Study of Work Teams, University of North Texas, www.workteams.unt.edu/reports

Handy, C. (1999) *Understanding Organizations*, 4th edn. Penguin Books Ltd, London.

Hawking, S. (2001) *The Universe in a Nutshell*, Bantam Press, London.

Hill, D.R. (1996) *A History of Engineering in Classical and Medieval Times*, Routledge, London.

Hughes, W.P. (1989) Identifying the environments of construction projects. In: *Construction Management and Economics* (7), pp. 29–40, E. & F.N. Spon Ltd, London.

Hutchins, T. (2001) *Unconstrained Organisations*, Thomas Telford Publishing, London.

Kalpakjian, S. & Schmid, S.R. (2001*) Manufacturing Engineering and Technology*, Prentice Hall Inc., New Jersey.

Kehoe, D.F. (1996) *The Fundamentals of Quality Management*, Chapman & Hall, London.

King, R. (2000) *Brunelleschi's Dome*, Pimlico, London.

LADC (1992) *The Skunkworks Approach to Aircraft Development, Production and Support*, Lockheed Aircraft Development Corporation, Burbank, CA.

Lawrence, P.R. & Lorsch, J.W. (1967) *Organisation and Environment*, Harvard University Press, Cambridge, MA.

Lawson, B. (1986) *How Designers Think*, Architectural Press Ltd, London.

References

Leitch, J. (2001) Hybrid forms take shape. In: *Contract Journal*, 24 October, Reed Business International.

Lientz, B.P. & Rea, K.P. (1999) *Project Management. Planning and implementation*, Harcourt Professional Publishing, San Diego.

Lipnack, J. & Stamps, J. (1997) *Virtual Teams Reaching Across Space, Time and Organizations With Technology*, J. Wiley & Sons Inc.

Luther, N. (2000) Integrity testing and job performance within high performance work teams: a short note. In: *Journal Of Business And Psychology*, 15, 1, Human Sciences Press, Inc.

Madine, V. & Black, S. (2001) The World is Not Enough. In: *Building*, 16 March, London.

Manz, C.C & Neck, C.P. (1995) Teamthink: beyond the groupthink syndrome in self-managing workteams. In: *Journal of Managerial Psychology*, 10, pp. 7–15.

Maturana, H. & Varela, F. (1980) *Autopoesis and Cognition: The realisation of The Living*, Reidl, London.

Maylor, H. (1996) *Project Management*, Pitman Publishing, London.

McCauley, J.L. (1995) *Chaos, Dynamics and Fractals*, Cambridge University Press, Cambridge.

McDiarmid, M. (1997) *Classic British Bikes*, Paragon Ltd, Bristol.

Meredith, J.R. & Mantel, S.J. (1995) *Project Management. A Managerial Approach*, 3rd edn. John Wiley & Sons Inc., New York.

Miller. E.J. & Rice, A.K. (1970) *Systems of Organisation: The Control of Task and Sentient Boundaries*, Tavistock Publications, London.

Mintzberg, H. (1979) *The Structure of Organisations*, Prentice Hall Inc., New Jersey.

Moore, D.R. (2000) Visual perception theories and communicating construction industry concepts. In: *Work Study*, 50, 2, pp. 58–62, MCB University Press, Bradford.

Moore, D.R. (2001) William of Sen to Bob the Builder: non-cognate cultural perceptions of constructors. In: *Engineering, Construction and Architectural Management*, 8, 3, pp. 177–184, Blackwell Science Ltd, Oxford.

Moore, D.R. & Ahmed, N. (1997) Proposals for the development of an indigenous materials and methods orientated design data aid for design professionals practising in developing nations. In: *Habitat International*, Sheffield, 21, 1, pp. 29–49, E. & F.N. Spon Ltd, London.

Moore, D.R. & Hague, D.J. (1999) *Building Production Management*, Pearson Education Ltd, Harlow.

Moore, T.A. & Moore, D.R. (1997) Project management for the construction industry: an initial examination of a systems approach, *ARCOM '97*, Cambridge, 15–17 September.

Mystrom, R., Baumeister, D. & Nerland, R. (1988) Anchorage Organising Committee for the 1994 Olympics. In: *Project Management Journal*, June.

Newcombe, R., Langford, D. & Fellows, R. (1990) *Construction Management: Organisational System*, Mitchell Ltd, London.

Osbourne, R.N. & Hunt, J.G. (1995) Environment and organisation effectiveness. In: *Administrative Sciences Quarterly*, June, pp. 231–246. Cited in Woodward (1996).

References

Parker, D, & Stacey, R. (1994) *Chaos, Management and Economics. The Implications of Non-Linear Thinking*, The Institute of Economic Affairs, London.

Petroski, H. (1996) *Invention by Design*, Harvard University Press, Cambridge, Massachusetts.

Pieters, G.R. & Young, D.W. (2000) *The Ever-Changing Organisation*, St Lucie Press, Boca Raton, USA.

Pilcher, R. (1992) *Principles of Construction Management*, 3rd edn. McGraw Hill, Maidenhead.

Pirsig, R.M. (1974) *Zen and the Art of Motorcycle Maintenance*. Corgi Books, London.

Reiss, G. (1993) *Project Management Demystified*, E. & F.N. Spon Ltd, London.

Sato, H. (1967) *Japanese Woodworking*, & Nakahara, Y. (1967) *Japanese Joinery*. Combined in *The Complete Japanese Joinery* (1995), Hartley & Marks, Vancouver.

Senge, P.R, Kleiner, A., Roberts, C., Ross, R., Roth, G. & Smith, B. (1999) *The Dance of Change*, Nicholas Brealey Publishing, London.

Shirazi, B., Langford, D.A. & Rowlinson, S.M. (1996) Organisational structures in the construction industry. In: *Construction Management and Economics* (14), pp. 199–212, E. & F.N. Spon Ltd, London.

Sorensen, R. (2000) The CM database: to buy or to build? http://www.stsc.hill.af.mil/crosstalk/2000/jan/sorensen.asp

Stacey, R. (1992) *Managing Chaos*, Kogan Page, London.

Theobald, R. (1970) *An Alternative Future For America II*, Swallow, Chicago. Cited in Banner and Gagne (1995).

Tichy, N.M. (1983) Managing organisational transformations. In: *Human Resource Management*, Spring/Summer, 22, 1/2.

Walker, A. (1996) *Project Management in Construction*, Blackwell Science Ltd, Oxford.

Walker, A. & Kalinowski, M. (1994) An anatomy of a Hong Kong project – organisation, environment and leadership. In: *Construction Management and Economics* (12), pp. 191–202, E. & F.N. Spon Ltd, London.

Ward, M. (1999) *Virtual Organisms*, Macmillan Publishers Ltd, London.

Wenger, E. (1998) *Communities of Practice. Learning, Meaning, and Identity*, Cambridge University Press, Cambridge.

Wilson, A.J. (1976) 13th century project management. In: *Building Technology and Management*, July/August, pp. 5–8.

Womack, J.P., Jones, T.J. & Roos, D. (1990) *The Machine That Changed The World*, Rawson Associates and Macmillan Publishing Company, New York.

Woodward, J. (1996) Organisational characteristics and technology. In: Cole, G. (ed.) *Management Theory and Practice*, 5th edn. DP Publications Ltd, London.

INDEX